"This book is extremely interesting and well written. Its logical progression of ideas culminates in a competently presented new metaphysic that deserves exposure in today's world.... The effect that this book has had on me, personally, is that it has refocused my inner eye, giving it X-ray vision to see deeper and clearer into the vast connectedness of everything that we so often take at face value."

—ANN MOORE
Winner, 2020 AVBOB Poetry Project

"In a world where philosophy is too often narrowly defined, *Everything, Briefly* is a work of amazing breadth and ambition. Presenting a view that is holistic, interconnected, relational, and boundless, it seeks to reshape our thinking about meaning, language, ethics, and religion, challenging many of our assumptions and setting out an entirely new metaphysics."

—**MEL THOMPSON**
Author of *The Art of Living*

"Thomas Scarborough's *Everything, Briefly* is wide-ranging and very engaging. It is refreshing—after so much work in analytic philosophy that is narrow in scope and highly technical—to see someone take on the project of constructing a panoramic metaphysics."

—**CHRISTIAN B. MILLER**
Author of *The Character Gap*

"What I most appreciate about this book is the recreation of a frame in which to have meaningful discussion around the connection of ideas and values. . . . Perhaps its most valuable contribution is the creation of a 'table' where all can come, sit, discuss, and identify what we have in common and brings us together rather than what divides us."

—**MAILIN YOUNG**
National Director of Leadership Development,
US and Europe InterVarsity

"This book will be a terrific help to all of us. Scarborough explains well our presence in and familiarity with the muddy physical reality, and then opens the door to the fuller reality of metaphysics, which we remembered, perhaps even craved, all along."

—**UDO MIDDELMANN**
Author of *The Innocence of God*

"Scarborough has provided an important resource material for those interested in metaphysics. It is new, fresh, and engaging. A must-read for the year!"

—**SAMUEL WAJE KUNHIYOP**
Author of *African Christian Theology*

"This metaphysic reveals the extraordinary intellectual investment Thomas Scarborough put into thinking deeply about the weighty issues of philosophy. The ambitious scope is striking; and it succeeds. The result is a profound, and often original, work worthy of joining the best from history. I'm impressed. His ideas offer much to ponder, appreciably presented in a delightfully accessible style."

—**KEITH TIDMAN**
Author of *The Operations Evaluation Group*

"*Everything, Briefly* is a unique metaphysic that seeks to unify nothing less than all human experience, and argues for unifying principles. There is breadth here, but not a removal from everyday experience. It's extraordinarily valuable to see beyond our immediate experiences. . . . As an antidote to tribalism and polarization, Scarborough's emphasis on connectedness is apt and timely—a starting point for conversations on the relatedness of all aspects of life."

—**STEPHEN GALANIS**
Philosophy Graduate, Rhodes University

"What are we doing here? *Everything, Briefly* gives us a new perspective on the journey that life is, starting with an examination of the basics—What is real?—then looking at language and its blind spots, and how that relates to our thinking; and taking us through ethics, science, mathematics, and the classics; and finally to places of deep meaning—religion and metaphysics. If you want to examine your life, *Everything, Briefly* can give you a place to start."

—**JILL HACKER**
NOVA Math Lab Tutor

"*Everything, Briefly* is a courageous attempt to subsume human experience and knowledge under a single metaphysics. Its largely impressionistic, bite-size style will appeal to the interested nonspecialist, while its unique perspective on some of the classic conundrums of traditional philosophy should pique the interest of academic philosophers of all stripes."

—**GEOFFREY MARNELL**
Author of *Correct English: Reality or Myth?*

Everything, Briefly

Everything, Briefly

A Postmodern Philosophy

Thomas O. Scarborough

Foreword by Mel Thompson

WIPF & STOCK · Eugene, Oregon

EVERYTHING, BRIEFLY
A Postmodern Philosophy

Copyright © 2022 Thomas O. Scarborough. All rights reserved. Except for brief quotations in critical publications or reviews, no part of this book may be reproduced in any manner without prior written permission from the publisher. Write: Permissions, Wipf and Stock Publishers, 199 W. 8th Ave., Suite 3, Eugene, OR 97401.

Wipf & Stock
An Imprint of Wipf and Stock Publishers
199 W. 8th Ave., Suite 3
Eugene, OR 97401

www.wipfandstock.com

PAPERBACK ISBN: 978-1-6667-3493-5
HARDCOVER ISBN: 978-1-6667-9146-4
EBOOK ISBN: 978-1-6667-9147-1

12/15/22

Cover art by Rachel Leibman

Internal illustrations by Thomas O. Scarborough

Scripture quotations are taken from the (NASB®) New American Standard Bible®, copyright © 1960, 1971, 1977, 1995, 2020 by The Lockman Foundation. Used by permission. All rights reserved. www.lockman.org.

Dedicated to the memory of Temeeti the son of Teiaa,
without whose influence this book would never have been written

Why do we admire the thinkers of Cretandom? We admire the thinkers of Cretandom because they have managed to reformulate the questions, usually in such a manner as to make certain answers at least seem plausible.

—SIR ISAIAH BERLIN

Contents

List of Illustrations | ix
Foreword by Mel Thompson | xi
Preface | xv
Abbreviations | xxi

Part I: Foundations

1 Low-Level Concepts | 3
2 A New Horizon | 9
3 What Is Really Real | 14
4 Limited Relations | 21
5 Liberating Relations | 28
6 Words and Relations | 38
7 Things and Relations | 45

Part II: Personal Ethics

8 An Endless Expanse | 53
9 Human Limitations | 62
10 Relations and Expectations | 71
11 Opaque Relations | 79
12 An Interrelated Ethics | 84
13 Esthetic Relations | 89
14 More about Words | 97

Part III: Public Ethics

15 Political Relations | 107
16 Informational Relations | 115
17 Employment Relations | 122
18 The Imposition of Relations | 129

19 Relating Identities | 135
20 A Boundless Environment | 143
21 Unvirtuous Behavior | 150
22 Critical Theory | 157
23 Awakening | 161

Part IV: Science and Math

24 Defining Ought | 169
25 Fact and Value | 174
26 Relating Science | 179
27 Relating Math | 189
28 Relating Languages | 197
29 The Isolation of Science | 203
30 Scientific Hubris | 210
31 Pseudo Ethics | 217

Part V: Classics

32 Deeper Relations | 225
33 Causal Relations | 232
34 Historical Relations | 241
35 Mind–Body Dualism | 247
36 The Purpose of Relations | 252
37 The Teaching of Relations | 258

Part VI: The Ultimate

38 Religion as Relations | 267
39 God beyond Relations | 275
40 Relations beyond Life | 285
41 Meaning as Relations | 291
42 What Is Metaphysics? | 297

Glossary | 305
Bibliography | 311
Subject Index | 331
Name Index | 349

List of Illustrations

1. A New Horizon | 12
2. Things are Concepts | 19
3. Subatomic Words | 31
4. Reality and Expectation | 73
5. Flows of Information | 117
6. Employment and Calling | 125
7. Levels of Abstraction | 202
8. Closed and Open Systems | 213
9. The Dissipated Word | 228
10. Mind and Reality | 250
11. Reason Bridges the Gaps | 255
12. Truth Claims | 274
13. Relations beyond Life | 288

Foreword

THE ASPIRATION THAT LIES behind this book is quite extraordinary and its publication timely.

Extraordinary, in that it seeks to breathe new life into metaphysics, the often-neglected attempt to grasp the nature of reality itself, rather than opting for the more narrowly defined questions that have become the norm in academic philosophy. Timely, in that it seeks to construct a theory based on relationships, interconnectedness and a holistic view, which reflects the global nature of the challenges, both intellectual and practical, that we face today.

Much academic philosophy, particularly in the analytic tradition, is precise and logical, making incremental steps within arguments that have been developed by generations of thinkers. Yet the real impetus in philosophy has always been provided by those who have not been afraid to challenge existing assumptions and raise fundamental questions.

Even an over-simplified and cursory glance at the landscape of Western thought, reveals the significance of Plato, with his theory of Forms and huge influence on both philosophy and religion, or Aristotle with a metaphysics that enables the integration of philosophy and the empirical quest of science, providing so many of the basic terms upon which later thinkers have come to depend, or Aquinas' attempt to integrate philosophy and religion. In the quest for certainty, and the shift in the 17th and 18th centuries towards the centrality of the theory of knowledge, you have Descartes, Hume or Kant, radically challenging previous arguments and setting the agenda for those who follow. You have the systematic thought of Hegel, or its political development in Marx, or the development of existentialism from Kierkegaard, or Heidegger's ontology. Hugely, for the 20th century, there is the influence of Wittgenstein and the shift in philosophy towards the appreciation of the limitations, meaning and function of language, or the pragmatist tradition of James or Dewey or the move towards postmodernism. These and other broad movements in Western thought have one thing in common; they seek to shift the philosophical agenda.

So what of Scarborough? His book is both challenging and disconcerting, in that it too seeks to shift the agenda. His metaphysics is systematically rooted in the ideas of interconnectedness and human values that come closer to the insights of the Buddha than to most western philosophy. He sets about using that new perspective in order to re-examine traditional areas of philosophical discussion – ethics, religion, free will and the relationship between mind and body.

Thus, for example, he sets out to reincorporate ethics into metaphysics, getting beyond Hume's claim that one cannot derive an 'ought' from an 'is' and thereby setting a question mark over much modern ethical discussion. His ten ethical principles are founded on the observation of reality in ways that bring it closer to the earlier Natural Law approach to ethics, and he is rightly critical of the false simplicity of much utilitarian thought. By thinking expansively and holistically, he is able to devise his own ethical maxims, taking us into intellectual territory far removed from the familiar arguments of utilitarian, deontological or virtue ethics.

In 1962, working in the field of the philosophy of science, Thomas Kuhn published *The Structure of Scientific Revolutions*. In it he described the long periods of time during which scientists tend to work within an accepted paradigm for their field of study, challenging it to a greater or lesser extent, until eventually it is found to be inadequate and is replaced in what he called a 'paradigm shift'. I sense that what Scarborough is seeking with this book is to present us with the need for a paradigm shift in metaphysics and in philosophy more generally.

Philosophy in the 20th century left few assumptions unchallenged, and its legacy is a suspicion of any 'metanarrative' or attempt to give an overall account of reality in general. This book attempts to rectify that situation and restore the quest to describe 'everything briefly' in a way that shifts our thinking from the confines of earlier paradigms to one that reflects the observation that there are boundless relations and interconnections within our world, and that our philosophy should build upon these 'low level' facts towards a genuinely new metaphysics.

I have little doubt that those steeped in academic philosophy may find some aspects of this book frustrating, partly because of the originality of what Scarborough sets out to do, and partly because the later sections of the book, in which he unpacks the implications of his thought, make one eager for further development and examination. This is not a criticism of the book itself, for it would be quite impossible in a single volume to unpack all the implications of his thought. What we have here is a work of genuine originality, shifting the parameters of debate and introducing ideas that may at first appear improbable because they are unexpected.

Perhaps there should be a final word of warning for those embarking on this book. Previous metaphysics has tended to argue for a single idea or theme around which everything else can cohere; a single key to unlock all mysteries. That is not the case here. In a world of boundless relations there is no fixed centre—nowhere to 'plant our flag', to use Scarborough's image—no narrow certainty, but only an invitation to an ever-wider sense of openness.

Those picking up this book, expecting it to offer a nuanced and careful development of familiar arguments, comfortably edging academic progress forward at a glacial pace, are in for a shock. It is not that sort of book, but it deserves to be read more than once in order to savour what it has to offer.

Mel Thompson
2022

Preface

ONE OF MY PROFESSORS sought to speak to a student about his worldview. The student responded, "Sir, I don't think we are communicating." My professor tried again. "Sir," said the young man, "I don't think we are communicating." My professor noticed that the student had prepared him some tea. He said gruffly, "Give me some tea!" Taken aback, the young man poured him some tea. "Sir," said my professor, "I think we are communicating."[1]

The curious thing about this exchange is that there was no communication at the higher level of abstract concepts, yet full communication at the lower level of a simple task: pouring tea.

If only it were so with philosophy.

And yet it *is so* with philosophy.

Traditionally, philosophy has traded in high-level concepts, which are broader and more complex than those that we find at a lower level.

Yet if we recast philosophy in terms of low-level concepts, it begins to look more like a conversation about tea. It becomes possible to discuss philosophical concepts without confusion. This is the method of the present book—without abandoning high-level concepts, which are retained for context.

A metaphysics is a total philosophy, a panoramic philosophy, which seeks to integrate all aspects of our knowledge and experience. The authors of the *Collins Dictionary of Philosophy*, Godfrey Vesey and

---

1. Schaeffer, *God Who Is There*, 96.

Paul Foulkes, wrote, "The enquiry is very general, and in a sense has to underpin everything else in the world."[2]

Of course, to write about *everything* is, in a sense, impossible. It may take years to master the most basic things: say geometry, boatbuilding, or poetry. Yet in the midst of an ocean of knowledge, there are some time-honored subjects, which, at various times, have been thought to hold the key to the unification of everything: among them, ontology; epistemology; axiology; and, more recently, linguistics.[3]

> Where is the center of events, the common
> standpoint around which they revolve
> and which gives them cohesion?
>
> —HERMANN HESSE

In this book, we find that the age-old subjects still deliver—if we drive the discussion down to a low level, to the nuts and bolts of thought, as it were. I part, therefore, with the pessimists. I believe that a new, all-encompassing philosophy is possible in our time. In fact, I seek here to demonstrate it.

A single historical development has made a new attempt possible. There has been a great shift in our conception of the *relatedness* of all things—influenced in turn by the advance of modern science.

This has introduced a new situation, in which we are more aware of an *interconnected* world than ever before. It has brought us, too, to the point where the solutions to our greatest problems lie well within reach. In all of our troubles and apparent stumbling in the dark, we have been edging ever closer to the answers. They lie in the observation of "the first scientist," Leonardo da Vinci: "Realize that everything connects to everything else."[4]

---

2. Vesey and Foulkes, "Metaphysics," in *COLDP*, 192.

3. The aim of metaphysics is, in my own definition, the integration of all fields of knowledge and experience, even if sub-fields should not receive specific treatment. For example, optics may be covered by the exact sciences.

4. Quoted in Fortuna et al., *Five Hundred Years*, 9. Bearing in mind that Da Vinci's "everything" was more constrained than our "everything" today.

Preface xvii

Development of the Book

I formed the vision for a new metaphysics during my first year in theological seminary, in Europe, in 1978. It was a time of great ideological divides—politically, socially, and philosophically—and there was growing relativism and pluralism, as had seldom been the case in the past. This provided a rich and lively academic climate.

I soon began to keep philosophical notebooks, which grew to thousands of entries—and soon I made trial runs on a metaphysics.

I found, however, that *integration* failed. I could not find the sense of cohesion I was seeking. Curiously, as I look back now, the cohesion is plain to see in my notes.

In 2004 and 2006, I wrote two short metaphysics for professors in the USA. But again, I felt that integration failed, although I obtained good grades. Finally, in 2008, holed up one icy cold winter in South Africa's Koue Bokkeveld to study and write, I felt that I achieved the breakthrough.

My core ideas were published in the Philosophical Society of England's centenary special edition of *The Philosopher*, in 2013,[5] and the year after this, the philosophy weekly *Philosophical Investigations* invited me to explore the wider implications of my thought. A constellation of essays now came into being, which tested the waters of a new, integrated scheme.

Editors Dr. Martin Cohen and Pierre-Alain Gouanvic raised the idea of a *metaphysics* with me and created an online space for me to develop the project. I completed this work in 2015, and it was published the same year under the title "Metaphysical Notes."

It was a small metaphysics, somewhat flawed, and taxing on the mind. Therefore, I embarked on a book-length manuscript in simpler style. As I did so, I had the privilege of wide-ranging comment from various thinkers, who continually lifted the level of my work.

Influences

My influences were many. Here I survey the most important few, both in my own immediate experience and in the philosophical past.

My parents grew up in different countries, under different powers. The tension between their two worldviews, their incisive thinking, and

---

5. Scarborough, "Revisiting Aristotle's Noun."

their broad learning contributed greatly to a rich thought life. My father, who was a minister, was an unconventional, inventive man, with a brilliant ability to sum up vast situations in a flash. My mother, not long after my birth, found herself seated opposite a translator for the United Nations on a train. He was intrigued to see her traveling with learned books in various languages and invited her that same day to serve as his assistant.

When I was just four years old, we moved to remote islands in the Pacific, where, apart from a few common utensils and cloth, Western influence was virtually unknown. There, a local man, Temeeti Teiaa, was appointed as my guardian. Temeeti, in an old world society, through countless activities over years, showed me a different way of life—and a different way of seeing things. He imbued me with a sense that his civilization was superior to mine.

I attended theological seminaries on three continents and had world-class professors. If I should single out any one of them, it was the philosophical theologian Dr. Francis Schaeffer, whom I quoted at the beginning of this preface. While I have carried few of his ideas into this book, he influenced me greatly through his emphasis on presuppositions and a holistic worldview.

I married a Swiss wife, Mirjam Rahel Meier, in 1982. She became a doctor of philosophy and a co-director and executive editor of the International Institute for Religious Freedom. She understood my ideas and dropped many passing comments. She died too young, of bone marrow cancer. Yet before she died, she followed a little-known tradition of the church, instructing me to marry Ester Sizani, a woman of largely Xhosa descent and royal lineage. Ester quietly yet powerfully promoted the completion of this book.

I received critical help besides, for which a few deserve special mention: a tutor in linguistics at the University of Cape Town, Atikonda A. Mtenje (now Dr. Mtenje-Mkochi); an academic supervisor at the South African Theological Seminary, Dr. Vincent Atterbury; a Rhodes University philosophy graduate, Stephen Galanis; an award-winning poet, B. Ann Moore (my sister); a psychologist and artist, Jeremy L. Dyer; a Swiss banker, Christian Maag; an insightful and combative editor, Dr. Geoffrey Marnell; a perceptive publisher, Wendy K. Lochner; and, again, Dr. Martin Cohen came to my side, bringing a maturity to the whole that I could not have achieved alone

Finally, as my manuscript edged ever closer to acceptance, several agents and publishers wrote to me along the lines that, while this was a fine work, they were not capable of it. It was only when I saw the enormous task it was for Wipf & Stock that I understood that the others had not made excuses. This book required extraordinary diligence and skill to bring it to fruition, over most of a year. I am especially thankful for Rachel Leibman's beautiful cover art, Mike Surber's spendid cover design, Dr. Rebecca Abbott's gracious and assiduous copy editing, Rachel Saunders' artistic typesetting, and managing editor Matthew Wimer's always timely guidance.

It is not easy to discern which *thinkers* were most formative in my quest. A few stand out in particular: Laozi, Nicolaus Copernicus, Søren Kierkegaard, Ferdinand de Saussure, Wilhelm Kamlah and Paul Lorenzen, Stephen Toulmin, Max Black, Jacques Derrida, Simon Blackburn, and Mel Thompson—and in a backward kind of way, the social optimists Karl Marx, Charles Sanders Peirce, and Pierre Teilhard de Chardin.

One writer stands out above all. I had read Aristotle once but found him to be all but inscrutable. I had made a two-page list of books that I thought to be essential reading for my task, and Aristotle's *Metaphysics* was last on the list. Now, he came alive. While I did not see in Aristotle what the commentators saw, his thought was catalytic for me. Reading Aristotle, everything fell into place.

A Note about Terms

While I leave any further observations about philosophy for the body of this book, there is something that seems to me to have a special place in the preface. Today there is a proliferation of *terms*, so that much the same things have many names. Take the example of a *thing*, a concept that is central to this book. A thing may also be called an object, entity, item, existent, being, concept, and so on—often with special nuances and conflicts of meaning.

Whatever the reason for it, one thing seems clear. If one dwells too much on the terms, one shall never write a metaphysics. My purpose is to create a harmonious whole, which is capable of uniting all knowledge and experience. While I carefully studied the terms, I finally chose those terms and meanings that would be useful to this end.

Thomas O. Scarborough
Cape Town

Abbreviations

CDP	*The Cambridge Dictionary of Philosophy*. Edited by Robert Audi. Cambridge, UK: Cambridge University Press, 1995.
CODL	*The Concise Oxford Dictionary of Linguistics*. 3rd ed. Edited by Peter H. Matthews. Oxford, UK: Oxford University Press, 2014.
COLDP	*The Collins Dictionary of Philosophy*. Edited by Godfrey Vesey and Paul Foulkes. London: Collins, 1990.
FDMT	*The Fontana Dictionary of Modern Thought*. Edited by Alan Bullock et al. New York: Fontana, 1988.
ODP	*The Oxford Dictionary of Philosophy*. By Simon Blackburn. Oxford, UK: Oxford University Press, 1996.
PDP	*The Penguin Dictionary of Philosophy*. Edited by Thomas Mautner. London: Penguin Reference, 2000.
SEP	*The Stanford Encyclopedia of Philosophy*. Edited by Edward N. Zalta. Stanford, CA: Stanford University, 1997–2021. Online resource. https://plato.stanford.edu/archives/.

Part I

Foundations

The purpose of this philosophy is not to supplant earlier philosophies but to unify and reconcile—so *incorporating* earlier philosophies in a new work. The first part of this book addresses the foundations of a new metaphysics.[1]

---

1. Foundations, not in the old sense of beliefs that require no further justification; rather, a *context* in which to develop a new metaphysics. Note here, at the very beginning, that there are some fundamental disagreements over the meaning of metaphysics. I define it as a grand unified philosophy, which integrates—or provides a *means* to integrate—all fields of knowledge and experience using familiar philosophical concepts.

Chapter 1

Low-Level Concepts

Philosophers have generally occupied themselves with high-level concepts. Yet solutions to philosophy's problems lie in low-level concepts.

IDEAS MATTER. THEY MATTER a very great deal. They matter to everything we see around us. They shape the world in which we live. This means that ideas are not merely interesting. More than that, they are of critical importance.

However, it is not those ideas which we *think* to be important that always are. This is especially true of philosophy. Take, for instance, individualism, *guanxi*, democracy, *laissez-faire*, or the scientific method. We may call these high-level concepts. They are big ideas, which, in general, enable us to better grasp our situation and explore it. Yet their track record is fairly poor, when it comes to solving the problems of our time.

There are, on the other hand, ideas of another kind, which are more important by far. Call them low-level concepts,[1] which are far more basic. They infiltrate our thinking everywhere. They are present in all of our speech and writing. Even before we begin to shape our high-level concepts, they are there. They existed where thought began.

As a result, what we see in the world around us may depend heavily on what might seem to be philosophical trifles: the *number* of things, things and relations, the opposites of things, and so on. These low-level

---

1. At the best of times, this is not a well-defined term. Here, it refers to concepts that lie at a level of abstraction below that which is normal or average.

concepts, I argue, are the real determinants of our thought life—therefore, our world, as we know it.

Consider an analogy from the world of computing. When we give a computer a set of instructions, we generally do so with a high-level language—feeding it lines that lie not too far from our everyday speech: INPUT receives input from the keyboard, ARRAY creates an array of data, WHILE executes a set of instructions while a condition is true, and so on.[2]

Yet it is possible to instruct computers more directly than this, through a language we call machine code. Machine code reads something like this:

```
01010011 01101111 01101100 01101001
00100000 01000100 01100101 01101111
00100000 01100111 01101100 01101111
01110010 01101001 01100001 00101110
```

Here, our instructions are reduced to the most basic unit of information—namely, individual bits. A bit can hold only one of two values[3] that correspond to the electrical values in the computer's logic gates: 0 or 1. As simple as it is, this code offers us supreme control over the machine.

Sometimes, there are errors or deficiencies at a low level, which a high-level language cannot address. Thus a program may not run, or it may be vulnerable to various kinds of failure. A high-level programmer may need to resort to machine code—alternatively, a low-level language—to overcome the problems that exist at a high level.

In philosophy, similarly, it is the simple things that are of supreme importance. Yet we have often overlooked them or neglected them. Or, we have let basic ideas slip into our minds without examining what they contain. In some cases, we have given them unquestioning acceptance for thousands of years, where they were not deserving of the honor.

The following examples merely illustrate—too briefly—the kinds of low-level concepts we shall examine more closely in due course:

- *Number.* How *many* things are there in this world? And if things, then how many relations between them? It is a critical question, which will become central to a new ethics[4]—yet it has been virtually

---

2. These particular instructions belong to the Pascal programming language.
3. This holds for a bit, though not for a qubit.
4. While ethics (singular) is a branch of philosophy, and ethics (singular or plural)

off the map for philosophers. For Socrates, it merely had to do with possessions: "How many things are there which I do not want!"[5]

- *Forms.* Plato, in the fourth century BC, held that there are abstract objects which, in the words of philosophy professor Sean McAleer, "are mind-independently real."[6] While we easily dismiss Plato's idea today, we retain it in all of our scientific theory, since theory is mind-independently real. It is not, we imagine, the product of fancy.
- *Things and relations.* Aristotle, in the fourth century BC, held the belief—still widely held today—that our whole world may be interpreted in terms of things and relations. Yet what are things? More than that, what are relations? This lies at the root of major philosophical problems.

While philosophers today have scrutinized many of the more basic concepts, a cursory look at a dictionary of philosophy shows that the ideas in which we generally trade—and in which we have traded for thousands of years—are predominantly high-level concepts from A to Z, from alienation to *Zeitgeist*—compatibilism, existentialism, irrealism, prescriptivism, reliabilism, or vitalism, to give but a few examples.

Conflict and Contradiction

Our failure to resolve philosophical problems at a low level, we shall show, has led to a breakdown at a high level and has obstructed our ability to unite all things—thus to understand the world in which we live.[7] Philosophy author Richard Osborne wrote, "It can be asked 'Where has philosophy and science got us to at the moment?' and the answer seems to be in a huge environmental and political mess."[8]

One does not need to understand much about philosophy to see the trouble. Both philosophies and philosophers have, from the

---

are moral principles, for reasons of style, we refer to both in the singular.

5. Socrates, as cited in Cicero, *Tusculan Disputations*, §32.

6. McAleer, *Plato's Republic*, 150–73. What Plato meant by forms is a subject of debate. His own understanding of forms may have changed during his lifetime. According to McAleer, Plato included mathematical objects under forms.

7. "The more banal the truth ('truism,' 'platitude')," wrote theologian Hans Küng, "the greater its certainty" (Küng, *On Being a Christian*, 74–75).

8. Osborne, *Philosophy for Beginners*, 181.

beginning, been in continual disagreement—and disagreement has resulted in conflict.

Every major philosophical tradition had its origins in conflict and *continued* in conflict. In China, it began with the Hundred Schools of Thought. In India there was, from the start, the separation between the orthodox and heterodox schools. The European tradition began with the great division that Diogenes Laërtius later called the Ionian and Italian—not to speak of the many divisions within divisions, which continue to this day.

This conflict includes the most fundamental subjects: the reality of our world, the existence of God, even the possibility of metaphysics itself. From the beginning, there has been an endless succession of philosophical contradictions and reversals.

Today, we may search in vain for a point of view that is not completely contradicted elsewhere. We see it in every dictionary of philosophy. Where we find realism, we find anti-realism; where we find theism, atheism; where we find altruism, egoism—and so on.[9]

A philosophical idea may seem to be persuasive for a time. The hard part is explaining why it and every other philosophical idea are so completely disputed and always have been.[10] The philosopher and activist Simone Weil concluded, "The contradictions the mind comes up against—these are the only realities."[11]

World-views may be grouped in pairs.

—EMIL BRUNNER

---

9. Classical skepticism held that the best methods deliver conflicting judgments about truth. The philosopher Francis H. Bradley held that contradictions are inherent in the basic categories (Bradley, *Appearance and Reality*, 518).

10. It is hard to know, wrote Georg W. F. Hegel, given so many competing yet internally coherent philosophies, which one is true (Hegel, *History of Philosophy*, 1:16–17).

11. Weil, *Gravity and Grace*, 151. The Madhyamika school of Buddhism taught that truth lies beyond all dichotomies and oppositions (Blackburn, "Madhyamika," in *ODP*, 287–88).

An Endless Cycle

Not only has there been *conceptual* confusion. Philosophers themselves have been in complete and continual conflict with each other.[12] This has been the case at least since Plato, around 400 BC. Diogenes of Sinope, on a visit to Plato's home, trampled on Plato's cushions with his muddy feet and said, "I trample on Plato's vainglory!" Plato replied, "How much pride you expose to view, Diogenes."[13]

In the Middle Ages, Abu al-Ghazali wrote the book *Self-Destruction of the Philosophers*—to which Averroës wrote the disparaging reply: *The Self-Destruction of the Self-Destruction*. Yet even in his own lifetime, Averroës was shunned. Sheikh Ash-Shuyookh Ibn Hamawiyyah summed it up: "He had many strange views."[14]

Centuries later, Voltaire, during the age of Enlightenment, wrote to Jean-Jacques Rousseau that, when he read his works, "a desire seizes us to walk on four paws."[15] Arthur Schopenhauer wrote that Hegel was an intellectual Caliban (the subhuman son of a malevolent witch);[16] Friedrich Nietzsche claimed that Schopenhauer was a case of the first order for a psychologist;[17] while Kant—a dignified man, who preferred to avoid personal insults—merely referred to his predecessors' *work*. He called it "obnoxious."[18]

At the same time, metaphysicians have been quick to extol the importance of their own ideas. René Descartes considered that he was rebuilding the house in which we all live;[19] Kant wrote that his was the only remaining path to truth;[20] and Ludwig Wittgenstein, in his first work, considered his thoughts "unassailable and definitive."[21]

---

12. The psychologist and philosopher William James reportedly said, "There is only one thing a philosopher can be relied upon to do, and that is to contradict other philosophers" (quoted in Mulvaney, *Classic Philosophical Questions*, 12).

13. Laertius, *Complete Works*, §26. Bertrand Russell gave an earlier example: Anaximander vs. Thales (Russell, *Our Knowledge*, 13).

14. Dhahabi, *Siyar Alaam an-Nubala*, 21:307–10.

15. Warner, "Voltaire to Rousseau," 38:15,484.

16. Jacquette, *Philosophy of Schopenhauer*, 79.

17. Nietzsche, *Twilight of the Idols*, 79–80.

18. Kant, *Critique of Pure Reason*, x.

19. Descartes, *Discourse on Method*, 32.

20. Kant, *Critique of Pure Reason*, xii.

21. Wittgenstein, *Tractatus*, 4. Ludwig Wittgenstein did not, however, think of himself as a metaphysician.

The problem with the philosophers of the past, wrote Bertrand Russell, was their "glib assertions [and] glib denials." But things had been "misconceived by all schools." He himself would now set it all to rights with "precision and certainty."[22] That attempt, titled *Our Knowledge of the External World*, is by now fairly much forgotten and certainly disputed.

The very fact that metaphysicians continually put each other down and set their own selves up—and in rare cases, put their own selves down again—ought to alert us to something that would seem to be larger than their very selves. Over and above these men and women, there is a continual cycle. One metaphysician displaces the other—or so they claim—only for another to displace them in turn—and this cycle does not end. The result is not only the confusion of the philosophers but the confusion of the world.

How, then, should we exit this scene, in order to see more clearly? The behavioral psychologist John B. Watson wrote, "Unless new issues arise which will give a foundation for a new philosophy, the world has seen its last great philosopher."[23]

---

22. Russell, *Our Knowledge*, 13.

23. Watson, *Ways of Behaviorism*, 14. Compare the philosopher and physicist Moritz Schlick, who considered that philosophy no longer has a domain of its own (Schlick, *Philosophical Papers*, 104). Theoretical physicists Stephen Hawking and Leonard Mlodinow wrote bluntly, "Philosophy is dead" (Hawking and Mlodinow, *Grand Design*, 5).

Chapter 2

A New Horizon

We live in a changed world, and a changed world gives us hope for a new metaphysics. Above all, we now see an endless expanse of things and relations.

UNTIL NOW, THE MANY conflicts and contradictions in philosophy have led us to a loss of *hope*—a dawning realization that we may never find an all-encompassing, unifying, harmonizing metaphysics. We have finally realized that, rather than advancing on the path of truth, we have merely been collecting worldviews grouped in pairs.[1]

Jean-François Lyotard wrote that there is an "incredulity toward metanarratives." We don't *believe* them any more.[2] The philosophical theologian Francis A. Schaeffer wrote, "There is a radical denial of the possibility of putting forth a circle which will encompass all."[3]

Today, it is said that the age of metaphysics is past. There is no possibility of a new metaphysics in our time. Every great tree in the forest has

---

1. Not that philosophy was in better condition previously. Immanuel Kant wrote, "At present . . . there reigns nought but weariness and complete indifferentism" (Kant, *Critique of Pure Reason*, ii).

2. Lyotard, *Postmodern Condition*, xxiv. This is rooted in turn, he wrote, in "the crisis of metaphysical philosophy."

3. Schaeffer, *God Who Is There*, 21. The philosopher Richard Osborne wrote, "The grand projects and absolute truths of philosophy have been crumbling away" (Osborne, *Philosophy for Beginners*, 181).

been cut down. We still find the remains on the forest floor, but it is plain to see that none still stands.

Not only this. Metaphysics has had awful consequences, historically. If metaphysics is about worldviews—which are fundamental orientations for living in this world—then people will die for them and kill for them.

Our metanarratives, wrote Lyotard, justify a single set of laws and stakes, which excludes all others.[4] If people are not moved by the right metaphysics, then this may be dangerous in the extreme. Rather, it is said, discourage metaphysics and focus on the smaller narratives of life, for which people will not die or kill.

This might seem to be a tempting thought, yet it causes us to lose sight of the great *significances* of life. Crucial issues of truth then pass us by—among them the total impact of our behaviors and the purpose of our civilization. To lose sight of such things carries with it its own great dangers.

In the twenty-first century, the loss of a compelling metanarrative has led to countless conflicts, which tear through our societies. People cannot agree about fundamental things: the nature of a constitution, the state of the environment, the seriousness of a pandemic, the meaning of violence, the role of the police, the desirability of free speech, even whether to eat meat or grow grass.

Science is what you know, philosophy is what you don't know.
—BERTRAND RUSSELL

A Changed World

Many people doubt today whether we can, *in principle*, describe everything briefly within one unified worldview. Even those who speak the same language may talk past each other. This idea was powerfully advocated by the philosopher Ludwig Wittgenstein, who proposed that we do

---

4. Lyotard, *Le Différend*, 143. Richard Osborne wrote, "Rather than telling truth, philosophy constructs meaning by suppressing, excluding, or marginalizing other terms" (Osborne, *Philosophy for Beginners*, 179).

A New Horizon

not in fact speak the same language—even if, as it were, we do.[5] Philosophy professor Milton Munitz summed it up: "Language is languages."[6]

Yet just as we seem to lose all possibility of uniting everything in a new, complete scheme, new issues have arisen. A major shift has taken place in our consciousness—and it has much to do with low-level concepts.

As our hopes for a new and unified account of reality have receded, so we have increasingly come to see the interrelatedness of all things. This has come about, above all, through our study of the lowest levels of the hierarchy of sciences: the *physical* sciences and chemistry, and their many and varied subfields—and with them, our increasing emphasis on data, logic, and math.

Through our advances in low-level concepts, it has become increasingly clear to us that all things are related to all things. Ultimately, there is nothing that is not related to everything else: not a word, not a motion, not an entity, not an event.[7] The philosopher of science Alan Chalmers put it like this: "Many kinds of processes are at work in the world around us, and they are all superimposed on, and interact with, each other."[8]

Through our advances at the lowest levels of the hierarchy of sciences, we have experienced a major shift in thought. Today, we have very much become aware of a new horizon, which is a practical infinity, not only of things but the relations between them.

New concepts signal the turn. *Holism* emphasizes the priority of the whole over the parts. *Emergence* speaks of order on a large scale, which arises from the simplest of entities. *Gaia* refers to a world in which all living beings are part of a single living being. Many buzzwords besides—among them, globalization, hypermobility, universalism, sustainability, environmentalism, internationalism, organicism, and the anthropocene—have to do with our rising awareness of an interconnected world and cosmos.

---

5. Where one cannot interpret one theory in terms of another, the two are said to be incommensurable—in general, called perspectivism.

6. Munitz, *Contemporary Analytic Philosophy*, 269–87.

7. While there may be no direct connection between, say, a bird in Hawaii and a fireplace in Tibet, the two exist only because they are related. The poet and printmaker William Blake wrote, "We see the world in a grain of sand" (Blake, *Poetry and Prose*, 493).

8. Chalmers, *What Is This Thing*, 26.

Part I: Foundations

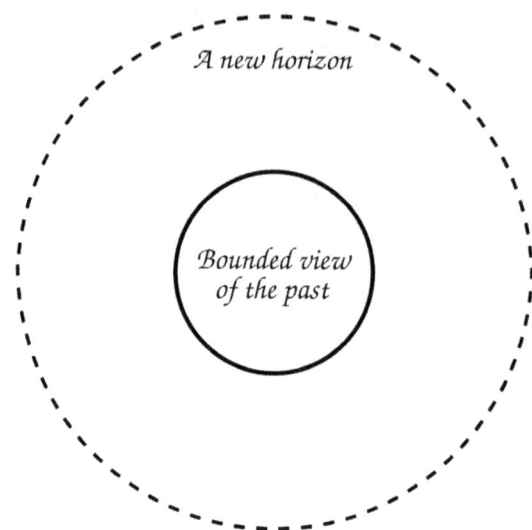

This stands in stark contrast with the past. Previously, philosophers had a *bounded* view of the world. They typically interested themselves in *immediate* things: for instance, the nature of matter, the requirements of the "good life," or the principles of politics—all of which were fairly close to home.

The ancient Greeks and Chinese, to be sure, had some appreciation of a total, integrated reality. Among other things, they wrote about the *logos*, the *tao*, panentheism, and the vital force. Yet these did not offer a shadow of the power that science has lent such concepts today. It is science that has shown us connections that the Buddhist expositor Lama Govinda called "an inseparable net of endless, mutually conditioned relations."[9]

To put it another way, without the explosion of knowledge at the lowest levels of the sciences, we would not have a new awareness of the interconnectedness of everything.

---

9. Quoted in Capra, *Tao of Physics*, 155. The semiotician Daniel Chandler interpreted literary theorist and philosopher Roland Barthes as saying that we have a "fundamentally *relational* understanding of reality" (Chandler, *Semiotics*, 125).

A Seamless Web

Imagine a mulberry tree, which stands in a field. Imagine that—as we sometimes do—we erase the tree's surroundings from our awareness and see only the tree. Yet gradually, we become aware that this tree stands in the sun, it moves in the breeze, its shadow falls on the field, and its roots reach into the ground. In short, we notice that this tree exists in *relation* with other things.

As we gaze upon the tree, we now think back in time. We consider the processes that formed the soil in which it grows, the rains that nurtured it, and the bird that carried its seed. We think to the future. We imagine mulberry pies, and moths that soon will fly where silkworms graze, and picnics in its shade.

Lifting our gaze beyond the tree, we see a radio mast, cirrus clouds, a flock of geese, the outline of a city, and a thin layer of smog. These introduce a blizzard of relations of all kinds, stretched out over times past, present, and future. We see spatial relations; temporal relations; causal, social, logical, and conceptual relations; and many more. The popular philosophy author Mel Thompson observed, "At any moment, we move within a seamless web of causality that goes forwards and backward in time and outwards in space."[10]

This is the new view—new, at least, in the sense that it has never before encompassed so much. Today, even those who are not philosophically minded are familiar in some way with our new horizon.

How, then, does this help us further? We shall see in due course that we are able to imagine not only an intimately interconnected world but, as we cast our mind across it, an endless expanse of relations—namely, relations between *things*. As we survey this expanse, we are able to discern certain *features* of the expanse.[11] These features, to borrow words from Descartes, are "clearly and distinctly perceived."[12]

On this basis, we shall be able to shape a new metaphysics, or total philosophy.

---

10. Thompson, *Philosophy of Religion*, 100.

11. The question arises as to whether we are *capable*, as finite beings, of appreciating an infinity or totality of relations. In a sense, no. Aristotle noted that "the infinite is potential, never actual" (Aristotle, *Metaphysics*, 50). Yet the works of the philosophers are replete with such concepts. I use the words "boundless" and "endless" in the sense of "all but infinite."

12. Descartes, *Meditations* (Veitch), 43.

Chapter 3

What Is Really Real

> How do we distinguish between real existents and useful fictions—or, for that matter, *useless* fictions? An important foundational question is, what is really real?

WE HAVE SAID THAT all things are related to all things.[1] "Everything connects to everything else." What, then, are things? What *kinds* of things are things? In fact, do things really exist? The question is not as absurd as it seems and is vitally important to later explorations of this metaphysics.

At a most basic level, we make a mistake when we think of things as real. Things exist in neuronal configurations in our brains, imprinted on a cellular structure.[2] Excise such things from the brain—say, a tree, a bicycle, or an apple—and we would know nothing of them.[3]

Therefore, the things that we know to exist, exist twice, as it were, and in two radically different forms. Those things that we are able to access immediately exist in the brain—but these things were previously converted from things out there, whatever those things may be. This naturally raises the questions as to what is in the world and to what extent we can know it.

---

1. Relations themselves may be called things—but we set this aside until later.

2. Compare Immanuel Kant, who held that, behind appearances (phenomena), lie things in themselves (noumena), of which we can know nothing (Stang, *Kant's Transcendental Idealism*, §6.1).

3. Not that this would be easy. Concepts and memories are to an extent *distributed* in the brain.

Even if we accept that all things are perfectly real, for the simple reason that they seem to be that way, there is little *agreement* today as to what is really real.[4]

Philosophy professor Simon Blackburn noted that "everything you can think of has at some time or another been declared to be a fiction by philosophers bent on keeping a firm check on reality—among them matter, force, energy, causes, physical laws, space, time, possibilities, numbers, infinity, selves, freedom of the will, the will itself, desires, beliefs, identity, things, properties, society, language, and money."[5] One may add, even metaphysics—and God.

One Kind of Real

When we examine the reality of things, the first thing that comes to our attention is that not all things quite seem to be *things*. Philosophy professors Bradley Rettler and Andrew M. Bailey separated things from things like this: "It is a sensible view that the concrete things are objects, and the abstracta are otherwise."[6]

Thus on the one hand, we have things like coffee cups and apricots and diamonds, while on the other hand, we have things like education and ownership and optimism. To put it too simply, there are things that exist in space and in time (we call them "ontic"), and there are things that apparently do not.

Even so, we curiously treat all things the same—or nearly so. When someone says that they had a happy childhood, that they have confidence for the future, or that they have many good friends, they consider that the things to which they refer are real. In fact, if we suggested otherwise—say, "friends aren't real" (they are abstracta)—they would surely think this very strange.

This immediately presents us with something of a crisis. There are a great many things that are not concrete, and we refer to them all the time. In fact, they are heavily represented in our one hundred most common English nouns: time, idea, force, and so on. We seldom question

---

4. It is very difficult, wrote the philosopher Aaron Sloman, to distinguish between real existents and useful fictions (Sloman, *Computer Revolution*, §2.4.2).

5. Blackburn, *Truth*, 2006.

6. The philosopher Gideon Rosen wrote, "It is widely supposed that every entity falls into one of two categories: Some are concrete; the rest abstract" (Falguera et al., "Abstract Objects," §1).

them[7]—and curiously, when we do, we question *some* of them more than we do others.

"Thing"... is applicable to any item whose existence is acknowledged by a system of ontology, whether that item be particular, universal, abstract, or concrete.

—JONATHAN ROWE

When we think of *concrete* things, we believe intuitively that they are real—since we see, hear, smell, taste, touch, or otherwise perceive them with our senses. Yet what are we to make of those things we do *not* perceive—either because, momentarily, we find that they lie beyond our senses, or because they are *constructs*—compound ideas, which may be lacking immediate empirical evidence?

Take the simplest distinction we make, between "real" things that we experience directly (say, trees in front of us), and "real" things we do not (trees behind us).

Imagine that I am cycling down a narrow bicycle track. In front of me, I see coconut palms—but the track behind me, I now no *longer* see—nor do I see the breadfruit trees I just passed beneath. However, I know that they are *there*. I saw them, smelled them, and heard their leaves rustling in the breeze. Besides, if I wanted to, I could stop my bicycle and look back to confirm it. In what sense, then, are those breadfruit trees *there*, which I no longer directly perceive?

This question received serious consideration in the eighteenth century.[8] How should I know that those things do not *vanish* which I do not see? This question was caricatured in verse by the theologian Ronald Knox:

---

7. We typically conflate words with things, wrote the professor of English Samuel I. Hayakawa (Hayakawa, *Language*, 30). This was all the more so in the Old World. It is reflected, for example, in the Eskimo proverb, "It is said that it is so and therefore it is so." (Pettazzoni, *Essays*, 16).

8. It had its origin in the philosopher George Berkeley, whose thought is summed up in the dictum "To be is to be perceived" (Berkeley, *Treatise*, 12). That is to say, something *ceases* to be when it is no longer perceived.

There once was a man who said: "God
Must think it exceedingly odd
If he finds that this tree
Continues to be
When there's no one about in the Quad."[9]

We may conduct a simple thought experiment.

Imagine that, as I approach the coconut palms in front of me, we switch off my senses and freeze this moment in time. Without my senses, the perceived and the unperceived now look largely the same in my mind. In fact, when one asks me about my reality at this point, I report that I know there still are coconut trees in front of me, just as there still are breadfruit trees behind me—all of them real.

Therefore, in my *mind*, there is little difference between the seen and the unseen—the heard and the unheard, the felt and the unfelt, and so on. Both exist in the vast neural network that is or contains the mind. Everything ends up there. The old idea that a construct, unlike a concrete thing, depends upon a subject's mind, is unsustainable. *Everything* depends upon a subject's mind.[10]

We need to be clear about what is intended here. From moment to moment, we do not deal with the *world*. We deal with the contents of our minds—and these are the world translated or reconstituted.

This observation now becomes important for the more vexing question as to how we are to understand abstracta.

There is a lot more to riding my bicycle than what I see, hear, smell, taste, and touch. If there were not, I would be lost on my bicycle, without anything to orient me. Do I have a destination? Should I turn back now? Do I *own* this bicycle? Do I need a passport here? Do I have a purpose here?—and so on. None of these surplus things is immediately real to me, yet I cannot afford to ignore any one of them.

My mind is filled not only with the things that I see or saw a moment ago, but it is filled with many things that are, in a sense, unreal. One could say, things that lack empirical evidence—although, in every case, they can be tested in some way.

Are these things real?

---

9. Ratcliffe, *Concise Oxford Dictionary*, 11.49.

10. What, then, of things that many people assume to exist, including *religious* objects so-called, where there is no empirical evidence? If the concepts exist, things become real as a result.

In fact, they are real—or as real as the coconut trees in front of me and the breadfruit trees behind me. This is because, together with the trees, they take the same form in my mind—each a distinct concept, with a unique label. As such, they do not fundamentally differ from those things that "exist."[11]

Excursus

We have a problem, someone says. Would this not mean that something exists merely because we *think* it does? We know full well that we are capable, as humans, of thinking of *fictions*. The ether, say, or (supposedly) fairies or tokoloshes.

The separation of the real and the fictitious may, however, be fairly straightforward. Real things largely correspond with reality as we know it, while the fictions contradict it. As for *appearances*—for example, a mirage in the desert or the mere playback of a sound—we generally grant them a special status in our scheme. We *understand* that they are mere appearances.

We shall return to this shortly—above all, to consider whether we speak of reality or fiction when we speak about God. Is God less real than, say, ownership or purpose—or even coconut trees? On what basis should we decide?

Things and Words

All things, we have said—that is, all things that lie within our knowledge and experience—exist as neuronal configurations in the brain, imprinted on a cellular structure, each a distinct concept, with a unique label.[12] Therefore, all things are real. They are real, we should say, to each thinking subject.

---

11. In fact, the discussion seems pointless, since we trade in abstracta all the time, almost completely without question or doubt. Notice that this represents an epistemology. Other aspects of epistemology are found throughout this book.

12. Unique, that is, within a given lingual context such as English. Outside such a context, different labels abound, and so do variations on concepts.

This means that both concrete things and abstracta are real—and there are no different classes, no degrees, no *types* of real.[13] We find that we cannot draw a *distinction* between concrete things, on the one hand, and abstracta, on the other, except on the rather weak basis that, in some cases, our senses are switched on, and in others, they are switched off. Apart from this, things are fairly much the same in the brain.

We call this a "flat ontology." There is nothing that is more real than anything else. Caterpillars are real, friends are real, time is real. If we have a *name* for it, it is real.

Another way of putting this is that concepts are things, and things are concepts. A concept is no less real than a thing, and a thing is no less real than a concept. This is so important that we shall repeat it:

Concepts are things, and things are concepts.

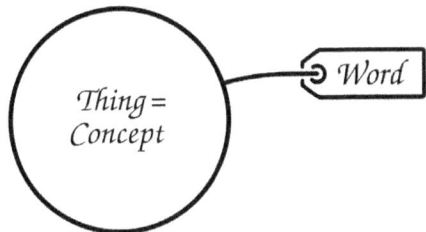

We turn now to the subject of *labels*. A label is a word or words by which we *identify* our concepts.

The linguist Ferdinand de Saussure held that we take a thing, and we put a name on it. The name is then united with that thing through an associative bond: "Each recalls the other."[14] Speak a name, and it recalls the thing; summon the thing, and it recalls the name.[15]

Philosophy professors Eric Margolis and Stephen Laurence put it like this: "Concepts are the meanings (or 'contents') of words and

---

13. Generally speaking, we believe that there are two kinds of things, namely real objects and abstracta. Some believe there are more. The philosopher Francis H. Bradley, as an example, divided things up into six categories.

14. Saussure, *Course in General Linguistics*, 78. Note that, while I follow Saussure's concepts, I do not use his terminology.

15. The philosopher and physician John Locke wrote, "Words in their primary or immediate Signification, stand for nothing, but the Ideas in the Mind of him that uses them" (Locke, *Essay Concerning Human Understanding*, 405). Even if we should call something a "thing." For example, "What is that thing over there?"

phrases."[16] A content is, of course, something that exists *inside* something else. We shall repeat this, too, so as to remember it.

Things are the contents of words.[17]

Now it might seem that the most interesting claim we have made so far is that *concepts are things*. Interesting, because we do not usually think of concepts in this way. We imagine that concepts are notional. However, this is not as important as the fact that *things are concepts*. If things are concepts, then things no longer have to be objects that, as software developer Ruth Ng put it, *cohere*.[18] A pot coheres, a person coheres, or a planet coheres.

When we think of things as concepts, they have far more possibilities than things which must cohere. Above all, they may be distributed or dissipated through space and time. They no longer need an outline, physical shape, or form. They do not merely exist on the inside of certain lines or bounds.

Such thoughts will surely seem unnatural to us now, but they will make sense in time. In later chapters, they help us solve a raft of philosophical problems. If there is any difference between a thing and a concept, then it lies in our pre-analytical understanding. When we call something a "thing," we imagine that it exists outside ourselves. When we call it a "concept," we imagine that it exists in the mind.[19]

We now need to note something of paramount importance to the chapters that follow. If things have labels, and these labels are words, then when we arrange words, we arrange things—that is, we associate things with other things in our minds. Therefore, we can learn a lot about the way that we arrange things by examining the way that we arrange words.[20]

---

16. Margolis and Laurence, Concepts, §1.3.

17. One might be tempted to say "nouns," not words. However, concepts are things, and things are concepts. This applies whether they are nouns or not. Let us replace the building materials of Wittgenstein's *Philosophical Investigations* with colors: "Person A is building with objects that are blue. Sky blue, baby blue, and angel blue. He calls out, 'Sky! Baby! Angel!' Person B brings the object which he has learnt to bring" (see Wittgenstein, *Philosophical Investigations*, §2).

18. Ng, *What Is an Object?*, question 3.

19. What reality itself is, is of little consequence to this metaphysics. We have everything we need in the concepts and things themselves, to take it from there.

20. There is an interesting correlation between the word frequency of "words" and of "things," which closely track one other. See, for example, Google's Ngram Viewer.

Chapter 4

Limited Relations

Metaphysicians have ambitions to think in every direction. Yet as they do, their thoughts are controlled by an unacknowledged metaphysics.

PHILOSOPHERS OF THE PAST, we have said, related things to things in their minds, yet they did so at a high level. Thus they ran into contradiction after contradiction—and conflict. Their failure to deal with simple assumptions at a low level caused conceptual obstructions in sometimes critical areas. These obstructions were deeply lodged in their thinking and remain largely, though not completely, unexamined today.

This is true not only of the metaphysicians of the past. Just as those philosophers were captive to contradictory ideas, so are we all. "We are all our own metaphysicians," wrote Godfrey Vesey and Paul Foulkes. "We all have our first philosophies."[1]

Just as the metaphysicians could not agree about things, neither can the rest of us.[2] Every day, we meet people who differ from us, not merely in the details but in their deepest thoughts, feelings, and actions— and precisely in those areas where we feel we ought to find our deepest sympathy, such as religion, politics, and ethics. While we all agree that "one plus one equals two," or "hydrogen and oxygen make water," when

---

1. Vesey and Foulkes, "Metaphysics," in COLDP, 52–53. Some refer to a *conceptual scheme*.

2. Aristotle noted, "For many men hold beliefs in which they conflict with one another" (Aristotle, *Metaphysics*, bk. 4, pt. 5).

it comes to such crucial subjects as reality, society, and God, we find that our conclusions are poles apart.

The time is ripe to enter upon some of the low-level concepts that matter to philosophical progress—and not only to philosophical progress but to various difficulties of the real world.

We all use *words*—and with words, we have said, we relate things to things, because things are the contents of words.[3] As we do this, we feel that we have the freedom to *arrange* these things with endless creativity. We call it the productivity of thought. Cognitive scientists Jerry Fodor and Zenon Pylyshyn wrote that human beings can think "denumerably many distinct thoughts" about things—which is to say, without number, although not without limits.[4]

We find the same idea in the field of linguistics, or theory of language. We feel that we have the freedom to arrange our words in any which way. In fact, we can create a potentially infinite number of sentences. We call this the productivity of language.[5] By so arranging our words, we become *creative*—turning out poetry, fiction, speeches, reports, and many other things—not least, philosophy.

But the feeling of freedom is a fatal mistake. Our words have a powerful hold on us. The ways in which we can arrange things in our minds are greatly limited. In fact, so powerful is the rule of words over our minds that this renders us fairly helpless in thinking anything that is new and objective. The literary critic and philosopher George Steiner wrote, "It is language that speaks, not, or not primordially, man."[6]

The Stubbornness of Concepts

We may think on the little pieces of colored glass that we use to create a mosaic—each of which is selected from a limited set of colors, a "finite

---

3. This differs fundamentally from the notion that things are the *referents* of words, as we shall see in due course.

4. Aizawa, "Productivity of Thought." There are certain thoughts we cannot entertain, such as a square circle.

5. The linguist Peter Matthews put it like this: "It is possible for any speaker to combine words into a sentence they have neither spoken nor heard before" (Matthews, "Productivity," in *CODL*, 321).

6. Steiner, *Martin Heidegger*, 22. The linguist Roman Jakobson pointed out, "Languages differ essentially in what they *must* convey and not in what they *may* convey" (Jakobson, "On Linguistic Aspects," 236).

stock."[7] The mosaic we create from this stock may represent anything we please—for example, ships on the sea or flowers on a table—yet the little pieces of colored glass are the most basic constituent parts that do not change.

Imagine that we stand back, to view the mosaic as a whole. All of its pieces in combination present us with a balanced and harmonious whole.

But consider what happens when we change a single piece in the mosaic—for example, a white piece on a white background for a black piece. We discover now that, if a single piece is out of place, this may be jarring to the eye—and a *few* such pieces out of place may give us the impression that our mosaic is *marred*.[8]

Therefore, while we have the freedom to create any picture we please, paradoxically, we may not have the freedom to put even one little piece of colored glass out of place.[9]

Now let us apply this to words—and again, things. Each word that we use has its own special place in a conceptual scheme. Each thing, too, has its own place in the world. More than this, each word and each thing may combine—or not—with the next. In most cases, not.[10]

Think for a moment on a word—any word. Say, a maid. A maid carries a pail—but she cannot carry the moon. She cannot carry a thought, a kiss, a sum, or a dance. We smile at such suggestions. This, and countless examples like it, teach us that we are very limited in the way that we can talk sensibly about a maid, a pail, or anything else, for that matter.

If we fail to take note of the rules of the system—or its *conventions*, if one prefers—we end up with a muddle of thoughts, as we find in the well-known nursery rhyme:

> Hey diddle diddle,
> The cat and the fiddle,
> The cow jumped over the moon;

7. "Finite stock" is a term preferred by Jerry Fodor and Zenon Pylyshyn (Aizawa, "Productivity of Thought").

8. We may compare our use of words with the periodic table, which organizes the chemical elements, each in its own place. When we put this table to use, it determines just which elements we may combine with others and which not. Similarly, words, like the chemical elements, will or will not combine with each other.

9. Depending, of course, on the make-up and character of the art—but the analogy will be understood.

10. It stands to reason. Things are the contents of words, and things are found together. Valleys and hills, bicycles and wheels, bread and butter, sunshine and summer all naturally occur at the same place and time or in the same way.

The little dog laughed
To see such sport,
And the dish ran away with the spoon.[11]

Yet our bondage is more subtle than this. Our words—indeed, concepts or things—contain deep assumptions about reason, being, identity, eternity, and a host of things besides, which rule over the way that we combine them.[12] Francis Bacon wrote, "For men believe that their reason governs words; but it is also true that words react on the understanding; and this it is that has rendered philosophy and the sciences sophistical and inactive."[13]

Excursus

Someone might object. If we are in some kind of bondage when we put words to words, is this not self-defeating? How should we hope to arrive at a new *philosophy*, if we are not free to think as we please?

However, we shall see in due course that it is not as though we have no means of escape. Rather, our thoughts are channelled by the *system* of language that we choose—that is, the scheme that we impose upon things—and this may change.

But first, we outline the troubles that afflict our present understanding of language—and together with our understanding of language, our understanding of things.

---

11. Whitmore, *Mother Goose's Melody*, 20.

12. An example is the philosopher René Descartes, whose core dictum was "[I doubt, therefore] I think, therefore I am" (Descartes, *Discourse on Method*, 46). He clearly assumed that his "I" was not to be identified with the body. Wilhelm Kamlah and Paul Lorenzen give the example of the polar-contrary opposition, which addled Christian Platonism "from Augustine to Leibniz" (Kamlah and Lorenzen, *Logical Propaedeutic*, 61).

13. Bacon, *Works*, 8:86.

Limited Relations 25

An Unacknowledged Metaphysics

We may explain our bondage by a theory of language—a *cluster* of theories—we call structural linguistics.[14] By its broadest definition, this is the dominant view of language today.[15] Language is "a self-contained, self-regulating system, whose elements are defined by their relationship to the other elements"—and while different linguists may organize the system differently, the idea of a great system remains.[16]

The linguists Wilhelm Kamlah and Paul Lorenzen observed that, in a linguistic system, words (more exactly, predicators) "always stand in such a tightly woven nexus that any one tends to appear with others."[17] That is, words powerfully attract words—but only *certain* words. They wrote, "We are thoroughly dominated by an unacknowledged metaphysics."[18] Even before we start out on a metaphysics, we already *have* one, although we do not know it. Again, we repeat this for its importance:

> *We are thoroughly dominated by an unacknowledged metaphysics.*

The philosopher Jacques Derrida, in his own obscure way, brought this bondage to the fore. He observed that metaphysics "return to origins"[19]—which is to say, basic principles. To say much the same thing, philosophers *launch forth* from origins.[20] They begin with some central idea—the cogito, the human will, the pure ego, and so on. From then on,

---

14. Matthews, "Structural Linguistics," in *CODL*, 385. In the words of linguist David Crystal, "any approach to the analysis of language that pays explicit attention to the way in which linguistic features can be described in terms of structures and systems" (Crystal, *Dictionary*, 457).

15. By its broadest definition, it would include generative linguistics.

16. The linguist Peter Matthews noted, "Different authorities have defined 'structuralism,' both in general and in specific application to linguistics, in what are at first sight very different ways" (Matthews, "Structural Linguistics," in *CODL*, 385).

17. Kamlah and Lorenzen, *Logical Propaedeutic*, 37. The linguist John R. Firth called it *collocation*: the regular appearance of lexical items together (Seretan, *Syntax-Based Collocation Extraction*, 8).

18. Kamlah and Lorenzen, *Logical Propaedeutic*, 10. The philosopher Max Black wrote, "An entire metaphysics is embodied in a certain language" (Black, *Labyrinth of Language*, 94).

19. Derrida, *Limited Inc*, 93.

20. Many have seen this as an insurmountable obstacle to a new metaphysics. Metaphysics has always interpreted the world in terms of *centers*: for example, Plato's forms; Hegel's concrete, abstract, and absolute; or Nietzsche's will to power.

their thoughts are powerfully controlled by the way in which words fit to words—or not.[21]

> The largest part of conscious thinking must be considered an instinctual activity, even in the case of philosophical thinking.
>
> —FRIEDRICH NIETZSCHE

Metaphysicians of the past, therefore, in order to describe everything, had ambitions to think in every direction.[22] Yet as they did so, they obeyed the rules of their linguistic systems and were tightly bound by these rules. Therefore, as they sat down to describe everything briefly, words were attracted to words, much as magnets snap to magnets, and so linguistic systems produced philosophies.

One merely needs to posit a starting point, and snap-snap-snap, one has a metaphysics. To put it another way, metaphysics *self-assemble*. It happens through the very nature of words and their attractional forces.[23] The philosopher and logician Gottlob Frege described it as "the rule of words over the human mind."[24]

In fact, *philosophers* may be at greater risk than the rest of us. Typically, philosophers are more careful with words. Therefore, they have less room to maneuver when they arrange them. The philosopher's "vantage point outside of things," as philosopher and psychiatrist Karl Jaspers described it, is illusory.[25]

---

21. Georg W. F. Hegel noted that philosophy's presuppositions are too often unproven and controversial (Althaus, *Hegel*, 222). Søren Kierkegaard suggested that everything may be reduced to arbitrary principles—say, boredom—or so I interpret him (Kierkegaard, *Either/Or*, pt. 1, 286).

22. Simon Blackburn wrote that the topics of philosophy "are life, the universe, and everything" (Blackburn, *ODP*, vii).

23. Prayer may serve such a function. If one posits God in prayer, all of one's thoughts are reordered in terms of this central idea.

24. Frege, *Philosophy of Logic*, 14. The philosopher David Hume wrote, "It is evident that there is a principle of connexion between the different thoughts or ideas of the mind, and that, in their appearance to the memory or imagination, they introduce each other with a certain degree of method and regularity" (Hume, *Enquiry*, 12).

25. This is not to say that philosophers should have no wish to be precise; rather, that precision may not save them from error.

Limited Relations

> ### Excursus
>
> Yet if it were all a case of "snap-snap-snap," would not words snap together to create the same results every time? Would not things, too, order themselves in our minds in predictable ways? How, then, do we account for the appearance of *different* metaphysics? Does this not prove that we are free to think as we please?
>
> In the coming chapters, we shall consider more carefully why philosophers ponder the very same subject matter, only to arrive at different—and often *opposite*—conclusions.
>
> We shall explain it like this: the same *subject* of which we speak may carry opposites within—good and evil, static and dynamic, large and small, and so on. These opposites offer different starting points—and so create different metaphysics. But we are getting ahead of ourselves.

We all have experience of self-assembling systems. Drop an idea into the middle of a conversation—or a starting point, a kernel, an origin, whatever one may call it—and thoughts multiply around it.

We sit down with a group of people in a hotel lobby. Our talk revolves around the tasteful furnishings and elegant décor. Suddenly, I drop a comment that I am doing some fascinating research into elephants. From this, a string of conversation results, about elephants, which occupies the whole group for quite some time—until a concierge interrupts us with a message. I wonder, then, at my powers of influence.[26]

This reveals our natural habit of assembling systems. We relate this thing to that, and that thing to the next, all the while being dominated by the way that our mind makes us do it.

---

26. It is possible, of course, that the group has no interest in elephants. Yet even this shows us that the conversation is largely predetermined.

Chapter 5

Liberating Relations

Metaphysics' bondage to words suggests that words may have a subatomic existence. In this chapter, we first explore this possibility imaginatively, as a first step to liberation.

WORDS LEAD US INTO bondage. Things, too, lead us into bondage, as things recall names, and names recall things and arrange themselves in our minds. It is hazardous to philosophy to think otherwise—namely, that we are free to use words as we please. If we imagine that we are free where we are not, we arguably commit the greatest of errors.

Yet what is it with words that gives them such power over us? What is it with things? What is it that drives our thoughts down predetermined pathways? Why are we not at liberty to arrange them as we please? What is this unacknowledged metaphysics, which addles our thinking? Why, if it exists, has it not been plain to view?

Again, it comes down to something that exists at a low level. It has to do with another, most basic assumption about words.

We think of words as "minimum free forms."[1] These minimum free forms have minimum *meanings*—called *explicit* meanings, which we generally sum up in dictionaries with the fewest of words. We think of words (more exactly, linguistic signs) as the *atoms* of language, in the sense that they represent language's most basic, irreducible components.[2]

---

1. Bloomfield, *Language*, 72.
2. In fact, we think of things in the same way, although not as obviously as we do

There is nothing more basic than words. We do not find that there are smaller particles or processes of which they are made up. *Morphemes*, perhaps, but that does not really add much to the matter.³ Or we could point to words' defining properties or resemblances,⁴ but these are fairly much at a minimum, anyway.

Now the question is this: in our current understanding of words, where could an unacknowledged metaphysics hide?⁵ If it is true that words rule over the human mind, this metaphysics must be hiding somewhere.

There seems to be one obvious way to approach our problem. We could look *inside* of words, if indeed we could, to see whether there is anything happening there, which, unknown to us, drives our thinking.⁶

An analogy would help.

Imagine a pair of loaded dice. With one roll of the dice, we ought to turn up any number between two and twelve, with more or less the chance that we expect. But if our dice keep turning up, say, seven, and seven again, and again, and again, then there surely is something going on *inside* the dice. The numbers we see on the faces are not the real story.

Another Vision

Cast aside, for a moment, our common notion of words as the atoms of language or something of the sort. Imagine, instead, a linguistic system—a conceptual scheme—where we may zoom in closer than the atoms themselves, to discover a *subatomic* existence of words. We shall think imaginatively here and work on the detail in the following chapter.

---

of words. Bradley pointed out that there is no "solid singleness" to things (Bradley, *Appearance and Reality*, 212). We return to this in due course.

3. One thinks also of bundle theory, according to which objects are bundles of properties—but not many.

4. One speaks of necessary and sufficient conditions (NSC), which is typically contrasted with prototype theory (PT). According to linguist Sebastian Löbner, the differences are "not as radical as PT made it to appear" (Löbner, *Understanding Semantics*, 182–83).

5. Wilhelm Kamlah and Paul Lorenzen proposed that we should intensify what we already do with words, namely tighten up their meanings, to weed out the unacknowledged metaphysics inside. The software architect Hermann Bense pursued their ideas and abandoned the quest: "We think it will dure some time until we come up with a stressable system" (Bense, "About Predicator.Name," para. 4).

6. The theory of concepts seems more promising than linguistics. It entertains a greater complexity. Eric Margolis and Stephen Laurence wrote, "Many concepts are themselves complex entities" (Margolis and Laurence, "Concepts," §2).

Imagine that we could crack open a word and look inside. Imagine further that we would find a world of *relations* there—that is, everything to which we relate this word and the thing that it names—everything, in every way.

We crack open the word "stone."

In our minds, we relate a stone to rivers and beaches and footpaths. We relate it to the sound it makes when we drop it onto other stones or throw it into a pond. A stone is hard and cold—and sometimes it is hot. We find white stones, black stones, even red stones and blue stones. We think of granite stones, pumice, and basalt. There are smooth stones and jagged stones. We walk on stones, throw them, collect them, and so on.

Excursus

Notice that, by saying that words contain a world of relations inside, we are reversing something that we normally accept about language. Ordinarily, we say that the mind makes a generalization of a concept—say, a stone—by extracting similarities from numerous examples of it.

What we do here is to gather all personal knowledge of a thing into a word.[7] Where we might usually say, for example, "all stones are hard" (a generalization of the concept), here we say that a stone "is" hard, whether true of all stones, or true of only one.[8]

For the sake of a name, we shall call this the synotation of a word (from the Greek *syn*, "together with," and the Latin *notare*, "to mark").[9] A word's synotation brings together, under a single label, every relation in which, as far as we know, it may be involved.

Let us say, therefore, that we find all of these relations *inside* the word "stone." Strictly speaking, we do not find relations "inside" a stone

---

7. Compare Karl Marx, who wrote that the seemingly simple object "abounds in metaphysical and theological niceties" (Marx, *Capital*, 1:§4).

8. One may take the example of a friend: he is strong, weak, bright, dull, kind, cruel, and so on. All these characteristics may indwell this friend. Actor and writer Stephen Fry noted that, at one and the same time, "Hermes and Hera and Zeus live within us" (Fry, in Peterson, "Atheist in the Realm," 1:16:12).

9. This is indeed a hybrid word, which combines Greek and Latin parts. While this may not be for purists, it is common to combine such parts.

or "outside," or anywhere else for that matter, but the goal is to obtain a working model of the word. For the sake of this model, we shall say that words contain a world of relations inside,[10] where "relations" stands for the many things to which the contents of a word are related in our minds.

Now let us think on how we *combine* our words with others—and with that, things with things.[11] It is the beginning of our liberation from bondage.

Imagine that a word, rather than combining with other words according to the rules of a linguistic system,[12] may potentially combine with other words in any which way, across all the vast expanse of our language—*on condition* that the synotations of both words overlap in some way.[13] Consider the following examples:

A bathroom is a room with a bath—yet in our experience, it typically contains a toilet. Therefore, the question, "Where is the bathroom?" will likely find us a toilet—though not in every case. We use various euphemisms besides: the privy, the powder room, the facilities, even "the euphemism"—and for the British, the water closet or WC.

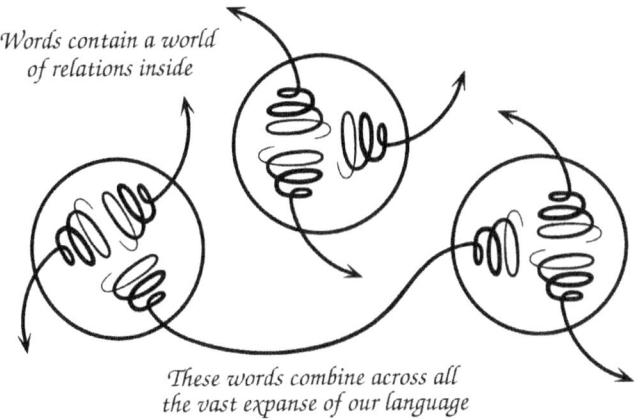

Words contain a world of relations inside

These words combine across all the vast expanse of our language

10. Interestingly, one may ask what a word *is*, which contains a world of relations inside. Is it anything at all? Wilhelm Kamlah and Paul Lorenzen wrote, "We state something about an object to which we point, in that we assert a predicator of the object." The object itself, without predicators, is apparently empty (Kamlah and Lorenzen, *Logical Propaedeutic*, 18).

11. Note that we are speaking about words, not (primarily) the reality to which they refer. Georg W. F. Hegel held that reality is constructed by the mind (Redding, *Hegel*, §3.3). In an important sense, this is true of the view developed here.

12. Not that we actually adhere to the rules.

13. Anaxagoras is thought to have held that there are traces of every substance in every other substance (Cohen, "Anaxagoras," §8).

Each of these words and phrases, and many more, is related to the real thing—be it through the presence of a bath, seclusion, the powdering of one's nose, water, and so on. Therefore, the euphemism is instantly understood. We could even try something entirely new: "Where is the soap room?," for instance.

One finds a similar dynamic with the *metaphor*: a man is an ox, a giant, or a machine. We know that an ox is strong, a giant is large, a machine is relentless and efficient. The same may be true of a *man*—such that these words may be *applied* to a man. Again, the *synotations* of both words overlap.[14]

Likewise the *idiom*—where again we take the word "stone": "It sank like a stone," "She was stone-cold sober," or "He turned to stone." The same applies to various linguistic features besides: parables, myth, and fiction, to give but a few examples. All depend on worlds of relations inside.

With such a view of language, we now find that we can speak freely. Our language is unbound: "The party got out of hand. Next thing, there were blue flashing lights. I threw my joint in the pool and scrambled over the wall."[15]

On the one hand, we have been free to speak like this since time immemorial. On the other hand, it is hard to reconcile this with our current theory. The philosopher Michael Losonsky wrote, "Language as human activity and language as system remain distinct focal points despite various attempts to develop a unified view."[16]

No dictionary will tell us that parties get out of hand—or that *hands* have anything to do with it—that police use blue flashing lights, or that pools extinguish joints. Yet we instantly understand it, and it shows us how we think. Words, in the way that we actually use them, have to do with every thing to which we have ever related them.[17]

---

14. Physics researcher Douglas Hofstadter and psychology professor Emmanuel Sander proposed that our thinking is defined throughout by analogy-making—a theory that has some relation to the concepts of this book (Hofstadter and Sander, *Surfaces and Essences*, prologue).

15. There are, of course, bounds to being unbound. We cannot surrender our language to major disorder.

16. Losonsky, *Linguistic Turns*, xv.

17. Aristotle wrote that, whenever two or more things are related, one finds their properties in both of those things (the relata) (Aristotle, *Organon*, 2:630). One might worry that this presents us with a fanciful way of relating things to things. What matters, however, is that relations between things are far more flexible, far more complex, far more vast than we have allowed.

Liberating Relations

How does this help us further? It would be useful to put it in a picture.

Imagine a game of dominoes. Each domino is rectangular, and the end of each is marked with a number of pips (one to six) or is blank. During the course of the game, we match end to end: six to six, three to three, and so on. As the game stretches out over the board, we create a simple, branching structure. So it is with words—in the *received* view of words. Words combine to grow into sentences, paragraphs, and chapters, like the branches on a domino board.

But discard the rules of the game, and unnumbered possibilities emerge. Instead of putting dominoes end to end, we might stack them in a pile, build little houses with them, or stand them on their ends in a row and watch the domino effect. The dominoes now become far more than what they were when they were still bound in a prescribed structure.

With reference now to linguistic systems, we may explode the limits of words and systems. Above all, we may explode the possibilities of creating a new account of everything, briefly.

Our words are not the atoms we thought they were[18]—nor do they contain the things we thought they did as we shall see in due course. This is a simple yet vital idea. Each of hundreds of thousands of words in our language contains a world of relations inside—a *subatomic* world—and these relations may bind with relations inside other such words, anywhere and everywhere.

But it does not end here. With such a view of words, we have an *overview* of our language. No longer do we look at combinations of words on a small scale but on the vast scale of synotation. We may now search for wider features of language as a whole. This is not something we typically do, and it enables us to reconceive reality in ways that differ fundamentally from the metaphysics of the past.

Textbook Examples

We return to an important question about our *bondage* to words.

If, in the received view of words, it is all a case of "snap-snap-snap" as metaphysics self-assemble, how, then, do we account for *different*

---

18. Hofstadter and Sander wrote, "Although dictionaries give the impresssion of analyzing words all the way down to their very atoms, all they do in fact is graze their surfaces" (Hofstadter and Sander, *Surfaces and Essences*, 4).

metaphysics? If we are in bondage to the words that we speak, why do we produce diverse outlooks on the world? Is this not a token of our freedom?

We shall consider the simplest and most obvious case: dichotomous ideas that shoot off in diametrically opposed directions—the *pairs* of worldviews that we find in this world. We shall describe this in terms of two classical frameworks of language: one Western, one Eastern.

From the strain of binding opposites comes harmony.

—HERACLITUS

In the simplest of English school grammars, our language is said to contain two basic syntactic components, which we call "principal syntactic elements." These are the subject and the predicate.

Today, one says that the predicate "completes an idea about the subject"[19]—and the fact that it *completes* an idea about the subject means that the idea—which is typically represented by a single word—is stripped down to something less than a whole. In fact, it becomes a *minimum* thing.

Take, for example, "the woman [subject] is good [predicate]." Here, we take "the woman" as a bare-bones concept that means an "adult human female"—no more, no less. Now we add to this the fact that she is *good*, and we have expanded on the concept of a woman, through the use of a predicate.

Yet suppose that the predicate—"is good"—serves not to *expand* upon the subject but to *narrow it down*. What if the predicate should take boundless possibilities contained in "the woman" and *channel* them?[20] The woman, who (potentially) could be and do anything and everything

---

19. While the subject-predicate distinction is a very old one—appearing in ancient Indian and Greek philosophy (but arguably, not Chinese)—it was in the thirteenth century that the philosopher and logician William of Sherwood treated the predicate logically—holding that it ampliates or enlarges the subject (Uckelman, "William of Sherwood," §2.2).

20. Philosopher Alexander Bain held that all existential statements have complex subjects from which a predicate can be extracted (Bain, "Symbols for Complex Subjects," in *Deduction*, 195). Melodie Fox, an instructor in information studies, pointed out that (for example) the word "woman" may contain stereotypical views that a definition bypasses (M. Fox, "Prototype Theory," 152).

that a woman is and does—"is good." We have now isolated just one aspect of her anything and everything.[21]

"The woman," rather than *excluding* most of the things we could relate to the word, now *includes* them. In fact, "the woman" must include *opposites*, which can split off in different directions.

Imagine a teacher who tells her class, "Write me a story about a woman." Some write about a good woman, some about a bad woman.

Similarly, philosophers write about realism and anti-realism, theism and atheism, altruism and egoism, as they ponder the very same subject matter.[22]

Excursus

It is a new idea, that the essence of a word includes every possible aspect of the subject.[23] This forces the meaning of "essence"—in fact, to such an extent that we should look for another word.

In keeping with our definition of *synotation*, we shall call this the *synessence* of a thing—from the Greek *syn*, "together with," and the Latin *esse*, "to be."[24]

The synessence of a thing, therefore, brings together every possible aspect of what that thing may be and do.[25] This differs from *synotation*, in that synotation is a linguistic term, while synessence is a philosophical term.

---

21. Someone might object. "I know a good woman. She is simply good. The possibility of her being anything else is not there!" Yet even if this is true, the point is that a word represents the totality of its associations.

22. One speaks also of markedness, a concept introduced by linguist Roman Jakobson. Paired signifiers (such as man-woman) consist of marked and unmarked forms, where the marked form is implicitly negative (Jakobson, *To Honor Roman Jakobson*, 3:268).

23. The cognitive scientist Lawrence W. Barsalou considered that we access different properties of a thing in different contexts (Barsalou, "Context-Independent," 82). A thing also has properties which we do not access.

24. It is a form of anti-essentialism. Synessence may be said to represent the indefinite article "a." The definite article "the" is closer to the essence of a thing, as traditionally understood.

25. Philosophy professors Sungho Choi and Michael Fara wrote, "Many objects

Opposites, therefore, are not extras that we insert into sentences as predicates: "The woman *is good*" or "The woman *is bad*." Rather, these opposites *already* dwell inside the subject, as *possibilities* within the subject. They belong to the totality of that subject[26]—in a sense, to its *essence*.

It is *possible* that a woman is good; it is *possible* that she is bad. It belongs to her *synessence* that she can be either.[27] It is *not* possible, however, that she should be, say, kiln-fired, gaseous, or botanical. This truly lies beyond the possibilities of "the woman."

Jacques Derrida recognized the more complex nature of the subject, in a different kind of way. He noted that we develop a "priority thought" about it[28]—so placing the predicate "good" before "evil," "positive" before "negative," "pure" before "impure," and so on—and so we return to origins.[29]

This has much in common with *Eastern* thinking, specifically that of the ancient philosopher Lao Tzu.

Lao Tzu taught that there is *unity in opposites*. Good reveals evil, beauty reveals ugliness, high reveals low, long reveals short, and so on. He further held that such opposites cannot exist *in isolation*: there has to be something that unites them. Opposites have to be true of one and the same *subject*: "Two names emerge from a single origin."[30]

---

differ from one another with respect to their dispositions in virtue of their merely possible behaviors, and this is a mysterious way for objects to differ" (Choi and Fara, "Dispositions," §1).

26. Immanuel Kant proposed synthetic statements, in which the predicate is contained within the subject—in a sense, synthetic words (Blackburn, "Analytic/Synthetic," in *ODP*, 15).

27. If we remove a feature from the synessence of a woman, the woman may no longer be a woman. Consider a synessence by which a woman is only good and does not contain the possibility of being bad. In this case, we have a redemptrix or a superwoman but not a real woman.

28. Derrida apparently equated origins and priority thoughts (Derrida, *Limited Inc*, 93). Note that a priority thought will to some extent be culturally determined.

29. Although it is not described systematically here, a completely new theory of language is suggested. Language is not the assembly of basic linguistic elements (morphemes, words, and so on) according to certain rules (morphology, syntax, and so on). Rather, one begins with the whole world in one's mind, and language systematically reduces this whole. Take "This city is tidy." "City" reduces the whole world to small regions of the planet, "this" reduces these regions to just one among many, and "tidy" reduces one region to a single aspect of it, while "is" reduces the time period of which we speak.

30. Laozi, *Almost Paradise*, 13.

Thus we may use the word "good" to describe a woman, a book, or a policy. Similarly, we may use the word "bad." But we cannot speak about good or bad in isolation.[31] They have to *belong* to something. In other words, while our words—specifically, our subjects, or *synessences*—may seem to be perfectly simple, inside them, opposites dwell.[32]

Again, these opposites may drive us in different directions, even as we speak about the same subject—and this may offer us some explanation as to how one and the same language results in diverse outlooks on the world.

In fact, there may be a *danger* in thinking in the old familiar way, of subject and predicate. If we discount a subject's predicates too soon or cast away its indwelling opposites, we may overlook what we have in front of us. We may meet situations without being open to their possibilities.

A Chernobyl becomes (merely) a power generator, a *Titanic* becomes (merely) a vessel that floats, a Putin becomes (merely) a head of state.[33] Words, stripped of their inherent possibilities, may dangerously limit our thinking and blind us to what happens next: a nuclear accident, a sinking ship, a delicate man, or a dictator as the head of state.[34]

---

31. One may ask what being and non-being are predicates of. An intuitive answer would be God. Compare the philosophical theologian Paul Tillich: "The question of the existence of God can be neither asked nor answered. If asked, it is a question about that which by its very nature is above existence" (Tillich, *Reason and Revelation*, 237).

32. Or *oppositions*, which include antonyms, directional opposites, complementaries, heteronyms, and converses.

33. Black, White, conservative, liberal, junior, senior, eastern, western, and countless terms besides do not merely have connotations but carry opposites within.

34. In the famous and much contested example of linguist and engineer Benjamin L. Whorf, empty drums explode (Whorf, *Language, Thought, and Reality*, 135).

Chapter 6

Words and Relations

*We now explore in detail the subatomic existence of words,
as a first step to regaining control of metaphysics.*

WE HAVE SAID THAT words contain a world of relations inside—and with them, things, because things are the contents of words. So far, however, we have been rather vague about it—such that we may be suspected of no more than speculation. Do words really have a subatomic existence? Do their synotations really reach across the universe?

This is where our foundational thoughts on ontology—"What is really real?"—have their first major application. It begins to matter that things are *concepts*, which do not simply cohere in space and in time. When we see that simple objects—call them concrete things or physical referents—are not what they seem, we find new ways of understanding our world.

On what *basis*, then, do we describe words as worlds of relations inside—rather than the minimum free forms with minimum meanings, as we have been taught? Are we able to *prove* that they are anything other than atomic?

Happily, we are in a position to reveal a lot about words *experimentally*. Here, we do it through a feature of language we call the *bridging inference*. A bridging inference ties two sentences together—yet rather than doing so explicitly, it does so implicitly, through the things that we *infer* about words—and things.[1]

---

1. While this applies to all parts of speech, we use nouns as examples.

Consider the two sentences "Aristotle owned a house. The plumbing leaked." The first sentence is about Aristotle's house; the second is about the plumbing. There is no explicit link between these two sentences, yet we *infer* that there is. We link them, because, intuitively, we understand that a house is related to plumbing—even if particular houses have no plumbing.[2]

Or, consider the two sentences "Aristotle owned a house. The karma was bad." Again, the first sentence is about Aristotle's house; the second is about something we find hard to put into words, which we call karma. Again, there is no explicit link between these two sentences, yet we infer that there is. We link them, because, intuitively, we understand that a house is related to karma.

This is not to say that we have deliberately made such links before, yet we are able to recognize them (almost) instantly, wherever we encounter a bridging inference.

We may list some further examples: "Aristotle owned a house . . ."

> The lounge was spacious.
> The view was amazing.
> The mist seeped through the cracks.
> The barracks were better.

These are just a few examples among many, and all reveal what we know could be true of a house.

The ease with which we make such inferences is all the more appreciated when we try some *incompatible* ones. For example, "Aristotle owned a house. The axle was broken," or "Aristotle owned a house. The preservative was vinegar." Intuitively, we know that such sentences do not belong together. In fact, it is astonishing how fast our minds are able to discover such incongruities.[3]

Much the same is true of a linguistic feature we call the *anaphora*, which is usually thought of as linking two sentences *explicitly*. Consider what happens when we convert our bridging inferences to anaphoras: "Aristotle owned a house. *It* had leaky plumbing," or "Aristotle owned

---

2. Here we speak again of synessence: the essence of a thing, which includes every possible aspect of that thing.

3. In linguistics, one calls it intuition. Daniel Dennett argued that most people instantly know that zebras in the wild don't wear overcoats, and countless facts besides, even though they have never stopped to consider them (Dennett, "A Cure for the Common Code," in *Brainstorms*, 104).

a house. *It* had bad karma." The anaphora—in both cases the pronoun "it"—is said to refer *explicitly* to the house.

Again we may add further examples: "Aristotle owned a house..."

> It was frequented by philosophers.
> It was looted by the mob.
> It was flattened by a meteor.
> It was blue.[4]

At first glance, there may seem to be no inferences here, only references. In each case, the anaphora "it" refers explicitly to the house—or so it is said.

Yet again, it becomes clear that we are dealing with inferences as soon as we try some false ones. For example, "Aristotle owned a house. It walked on all fours," or "Aristotle owned a house. It was poured from a teapot." We see here that the anaphora, too, has to be compatible with various inferences that we make about a house. Such inferences are a precondition for the anaphora to work.

Both the bridging inference and the anaphora—in fact, various features of language besides[5]—reveal that we make inferences that are far richer than the dictionary definition of a word and make mischief with the common notion of a word.[6]

> Every utterance, no matter how laboured,
> trails clouds of implication.
>
> —MAX BLACK

What we see, then, is that if there is any relation between two things—between a house, say, and its karma—or its chimney, its occupants, or its area in square yards,[7] these things all, inevitably, are wrapped up with the essence—we now say *synessence*—of a house. But if there is no relation—between a house and an axle, say, or between a house and its

---

[4]. With these and previous examples, we have linked the concept of a house with a large number of things: esthetics, sociology, cosmology, law, defense, and so on.

[5]. Including nouns, pronouns, verbs, adjectives, and adverbs.

[6]. It is a fact of language, although we may find it difficult to fit into existing theories.

[7]. Or, we may think of associations external to the house—say, a house's relation to other houses, cities, continents.

preservative—then these words will have little or nothing to do with each other. They will fail to combine when they come within each other's orbit.

Notice now what we have done. We find that the *atomic* view of the word is gone, and a *subatomic* word has taken its place—a word that is filled with relations inside. We are beginning to understand where an unacknowledged metaphysics may hide.

Incorporating Experience

Let us now take a brief detour to consider another dimension of words—one that was especially important to the later Ludwig Wittgenstein. Our words are not merely of academic interest. Many of the relations we find inside words are formed through the complexities of our experience of the world. This creates a direct link between our words and the reality that they infer.[8]

Although we may discover the meaning of words through a purely mental process—for instance, by consulting dictionaries or reading language courses—we absorb many relations of a word through our *experience*: say, through a dance, a sunset, or a storm.

Words have to do not only with every relation we have ever thought of but with every experience we have ever had. All of this is gathered into the synotation of a word, which is the synessence of a thing. Our words have everything to do with their *context* and, therefore, to use a term of Wittgenstein's, *the form of life* within which we use them.[9]

Consider this example.

Some friends drop by at my house after a hike. They tell me, "We didn't quite make it back by nightfall. We were still on the trail. But it's a full moon tonight." I instantly understand it—namely, that they were not lost in darkness. This is not because I read about the luminosity of the full moon, or because dictionaries include such information, but because I have experienced it myself.

Now if a word has a subatomic existence—since it is a world of relations inside—it follows that there can be no dividing line between denotation and connotation—or, for that matter, between what are called

---

8. "Infer" rather than "describe." We do not imply that the relations that we find inside a word are soundly mapped onto the world, or vice versa.

9. Wittgenstein, *Philosophical Investigations*, §19. Similarly, John L. Austin noted that speech acts require preparatory conditions (Searle, *Speech Acts*, 69).

its essential and accidental properties. We are left only with *synotation*, which is every relation in which the word is involved.

What, then, are we to do with *definitions*, if they are generally understood as the most economical statement of descriptive meaning?[10] And what of the distinction between denotation and connotation—where, to put it too simply, denotation is the dictionary definition, and connotation is (my preferred meaning) the *illusion* of denotation?

We cannot fit a whole world of relations into a dictionary definition. If we want to make a definition, it will have to be far less. Therefore, a definition that does not recognize a concept of denotation will need to describe *priority relations* inside a word.

In the case of a stone, a definition might be: "hard mineral matter, not metal" or something of the sort. While this does not differ essentially from the dictionary definition we know, it has something that dictionary definitions do not have. It is deliberately understood as prioritizing or reducing *relations* within the endless expanse.[11]

When we use words, then, we do far more than putting dominoes end to end. The big picture is that hundreds of thousands of words in our language contain a world of relations inside, which may bind with relations inside other words, everywhere. Rather than representing a finite structure, "a self-contained, self-regulating system," our language reveals boundless relations across an endless expanse.

We all knew it long ago, from the movies. A man walks into a bar and announces, "The sheriff is back in town." Men straighten up in their chairs. One chokes on his drink. Another drops his cigarette. Fear clouds their eyes. Yet no more has been spoken than "the sheriff." The power of the scene lies in the relations that a single word contains—which, in our mind, combine with many other relations inside other words.

With a new understanding of words (and with that, a new understanding of things), we begin to see countless relations—and inferences, implications, possibilities, corollaries—which we may well miss with the common understanding of words.

---

10. The linguist Geoffrey Leech wrote, "The analysis of word meanings is often seen as a process of breaking down the sense of a word into its minimal distinctive features" (Leech, *Semantics*, 96). Note, however, the qualifier "often."

11. We are seeking to describe definitions where denotation is rejected as a valid concept. Of course, many definitions include relations. Aristotle, as an example, wrote that a house has three essential features: stones, bricks, and rafters. These are "disposed in a certain way.... The whole is not a heap" (Aristotle, *Metaphysics*, 223).

The Truth of Words

There is an important consequence of thinking on words as worlds of relations inside. We could easily miss it. We typically think of words, we said, as the *atoms* of language. Then we combine these atoms to create a potentially infinite number of sentences.

We may now note that, in this received view of language, *sentences* may be true or false (we call it bivalence), yet *words* may not.[12] The linguist Geoffrey Leech wrote, "Truth and falsehood . . . are properties of sentence meanings but not of word meanings." Words in isolation are viewed as being *neutral*, with regard to truth.[13]

Yet all this changes, if we adopt the view that words have a world of relations inside. These relations may now be so arranged that they are true or false. In other words, words may be true or false *in themselves*, because they contain *sentences* within or something of the sort.[14]

This is usually hidden to us. By and large, we accept words uncritically. We do not deliberately consider their truth or falsity. While we constantly assess the truth or falsity of *sentences*, our words are generally accepted without question.[15]

The philosopher Cyril E. M. Joad was famous for the line "It all depends on what you *mean* by [a word]."[16] Through this, he indicated that our words may be false in themselves—at least, false in what we *think* they are in themselves. If we think we are speaking of science when we

---

12. This refers primarily to declarative sentences, but (arguably) to imperative, interrogative, and exclamatory sentences, too.

13. Leech, *Semantics*, 105. Thomas Mautner wrote that "'true' and 'false' do not apply to concepts or terms. It is sentences, statements, propositions, beliefs, theories and doctrines that can be said to be true or false" (Mautner, "Concept," in *PDP*, 103).

14. Aristotle wrote that a noun is "a sign of the definition," and "definition is one discursus or sentence" (Aristotle, *Metaphysics*, 223). He may simply have been stating that a noun is defined by a sentence or discursus, yet it would seem equally possible that he was saying that all the relations we find in a sentence or discursus may be wrapped up inside a word. Alfred North Whitehead may have suggested such a view when he wrote, "An old established metaphysical system gains a false air of adequate precision from the fact that its words and phrases have passed into current literature" (Whitehead, *Process and Reality*, 13).

15. The physicist Georg C. Lichtenberg similarly considered, "Our false philosophy is incorporated in the whole of language" (quoted in Cooper and Spolsky, *Influence of Language*, 11).

16. Joad's perennial catchphrase on the *BBC*'s The Brains Trust (Ratcliffe, *Concise Oxford Dictionary*, 10.26).

are really speaking of technology, or if we think we are speaking of average when we are really speaking of median, then our words are false in themselves.

Yet more than this, words may be false under all circumstances, in all contexts. We may ask some hard questions about certain words—such as "victory," "righteousness," or "altruism"—not to speak of religious objects so-called, such as "angels," "demons," or "God."

We are quick to say, "You lie!" when a sentence is false. Yet when we examine our words more closely, exploring their subatomic existence, many of the words that we speak may be a lie, and many more will apply to our reality only fairly loosely, so that they may easily deceive.

Can one truthfully speak of victory? There may be no true victors in this world. Can we truthfully speak of righteousness, where there is none righteous, not even one? Can we truthfully speak of altruism, where no such thing may exist? Yet we accept such concepts unquestioningly, because they elude all checks for truth or falsity.

Here conclude the details of language. Above all, the picture they provide of the relatedness of all things will help us to develop a new *ethics* in chapters 8 and 9. An expansive view of language will enable us to see new *features* of language—and of reality. These features have everything to do with our everyday living and promise to deliver us, in the realm of ethics, at least, from a morass of philosophical contradictions.

Chapter 7

Things and Relations

The number of relations we find in this world is *unbounded*. This simple insight is essential to the development of a new metaphysics.

OUR STUDY OF THE nature of words in the previous chapter shows us—reminds us—that we can relate things to other things far more expansively, intelligently, and imaginatively than we often do.

But so far, we have spoken about relations between things in fairly ordinary, everyday terms. We now consider the purely *quantitative* connectedness of things, which is a far more *abstract* connectedness, the bare *number* of things—and relations—in this world. It might seem trivial[1]— as to whether there are millions of things, quintillions, or perhaps an infinite number—yet it matters a great deal. There are core problems of philosophy we cannot solve without giving it some thought.

Most importantly, the number of things in this world—once we have assessed it—enables us to develop a concept of *totality*, which we do not gain by studying individual entities or fields of thought. In this way, above all, we may assess what powers we have in understanding and controlling our world.[2] This has everything to do with our behavior.

---

1. Physicist Richard O. Colestock summarized the attitude of many people towards the number of things: "What are you going to do with this answer once you have it?" (Colestock, "How Many Things?," para. 4). Simon Blackburn wrote, "But there is little useful that can be said about everything that is real" (Blackburn, "Being," in *ODP*, 40).

2. And other worlds, such as Mars, should we reach them.

Therefore, the simple, low-level concept of number becomes important not only for personal conduct but also for public ethics, environmental ethics, the pursuit of science, and many things besides.

Quantity of Relations

The question of the *number* of things has been largely overlooked. While philosophers have given much thought to things and *sorts* of things, they have had little interest in how *many* things there are, outside of any given field. Bertrand Russell asked—without any concern for number—"What *things* are there in the universe?"[3]

> When the understanding is once stored with . . .
> simple ideas, it has the power to repeat, compare, and
> unite them, even to an almost infinite variety.
>
> —JOHN LOCKE

The things that we relate to one another, firstly, are not *fixed* in number. We see this in our language, which adds and subtracts things all the time. Mostly, it adds things. Geoffrey Leech observed, "In a language like English, new concepts are introduced in large numbers day by day and week by week."[4]

Secondly, there is no end to the things we can create in our minds: for instance, clouds with noses, dogs that wag their tails, names that start with a J, ants that fall off the wall, and so on, indefinitely.[5]

While some such things may seem strange, they are nonetheless *things*, which we may relate to other things—even mathematically, if we

---

3. Russell's full sentence: "What things are there in the universe whose existence is known to us owing to our being acquainted with them?" (Russell, *Problems of Philosophy*, 92).

4. Leech, *Semantics*, 35. Much of this has to do with advances in knowledge.

5. John Locke held that all ideas of substances—animals, minerals, and plants—are complex ideas (Losonsky, *Linguistic Turns*, 9). The philosopher and mathematician Charles Sanders Peirce suggested tentatively, "A Sign may have more than one Object. The set of Objects may be regarded as making up one complex Object" (Rosensohn, *Phenomenology*, 96). This multiplies *things*.

please: "Ten ants fell off the wall, while seventy clung to it, so that 10 ÷ (10 + 70) x 100 = 12.5 percent of all the ants were (call them) fallers."

Thirdly, it is a simple fact that the number of things in this world is unthinkably large.

I stand on a hilltop and look down on a city. I see one hundred thousand homes—which is 10^5. An insect flies before my face. It is one of ten quintillion on the planet—or 10^{19}. I take a breath of fresh air. It contains ten sextillion molecules—or 10^{22}.

In fact, the *classes* of things alone (for example, mammals, musical instruments, or succulents) are unthinkably many, let alone their individual members.[6]

And then, apart from things in themselves, the *relations* between things are even more. A familiar example is the relations that exist between people in a room. If we think only of the pairs of people who might exchange glances, with three people, there may be three pairs; with six, fifteen; with nine, thirty-six; and so on.

Yet to count glances is to count only the simplest of relations. We could count those people who gave others the *first* glance or those who gave them a *downward* glance—or we might put glances in chronological order, and so on. We find reflexive relations, transitive relations, symmetrical relations, and many more. The number of possible relations now grows exponentially.[7]

We shall claim, therefore, that the total number of relations in this world is for all practical purposes infinite. The philosopher and physician John Locke used the phrase "almost infinite."[8] We shall use, instead, the word "boundless." We are not able to determine their bounds.

Again, not all things that we relate to one another are concrete objects like stones or insects or molecules. We may also speak about things that *happen* (events), things that exist in our *minds* (concepts), and various things besides[9]—all of which we may potentially relate to one another. We could, for instance, relate a battle to a war, deduction to induction, a week to a month, and so on, where none of these things are physical objects.

---

6. We do not use "classes" in the same context as ch. 3, where we said that there are no separate classes of concrete things and abstracta.

7. Relations may also be both real and notional. Notional, for example, if we blame beetles for eating our garden vegetables, where really it is caterpillars.

8. Locke, Essay Concerning Human Understanding, 119.

9. For example, agents, patterns, and expressions.

Further, the kinds of things we relate to one another may exist on different *planes*—something that we sometimes illustrate with the pyramid of sciences. We relate things mathematically, physically, chemically, physiologically, sociologically, psychologically, philosophically, theologically, and in many other ways—and on each of these planes, we relate things to one another in ways that are special to that plane.

Typically, we are unable to describe a physical concept (say, a proton) in psychological terms, a theological concept (faith) in chemical terms, or a mathematical concept (a square root) in physiological terms—unless, that is, we depart from our normal usage of words. Thus each field of knowledge represents a vast and interconnected web of relations in itself.

The Strangeness of Relations

But relations between things may be stranger than we think.

Consider an imaginary situation. A peal of thunder rolls across the city.

"It's related to the heat," I say.

"Oh? How so?" asks a friend.

"It's *always* like that," I reply.

Another friend offers, "Heat rises. Clouds form. Charges separate. That makes lightning—and thunder." My own answer now looks foolishly simple. Besides, what did I *mean* when I said that the thunder is related to the heat? How much did I really know?[10]

In a best case, when we talk about relations between things, we are using a kind of shorthand for something about which we are not being explicit. In a worst case, the missing links represent ignorance. I may be completely ignorant, for example, of what connects the heat with the thunder, although I might be right about the relation itself.[11]

Consider our imaginary example above. "Heat rises" is a thing that *happens* (an event); the same is true of "clouds form"; the same of "charges separate." Where at first I spoke about a relation, I find now that

---

10. The moral philosopher Stephen Toulmin noted that, if one predicts something on the basis of the regularities of nature, this is not to say that one understands what is happening (Toulmin, *Cosmopolis*, 117).

11. In fact, even when we speak of *fictional* relations—I might claim, for instance, that thunder is related to the phases of the moon—there remains a physical connection. The fact that we have related things to one another is inscribed in the physics of our brain.

there are multiple *things* in its place—but not only things. There are new relations there, too. Heat is related to the formation of clouds, the formation of clouds is related to the separation of charges, and so on. And when I examine these new relations, again, I find *things* in their place. It is an infinite regress. The philosopher Francis H. Bradley wrote, "We are forced to go on finding new relations without end."[12]

What, then, are relations?

Fraser MacBride, a professor of logic and metaphysics, noted that relations are "difficult to locate."[13] They are not substances; they are not attributes. What *are* they, then? Some deny their existence altogether.

For the time being, we shall stay with the simple view that things and relations are just what we think they are—but in due course, we shall see that they are far more *fluid* than that—in fact, *need* to be more fluid, if we are to solve certain philosophical problems.

Two concepts need clarification before we continue.

During the course of this book, we shall speak about *tracing* relations. This means *marking the course* of relations—pursuing them, following them, as we connect one thing with the next and the next.

We shall also speak about *regions* of relations. This refers to *webs* of relations of appreciable size and complexity—say, the relations between things that describe a particular field, such as economics or engineering, or the relations we trace when we form an opinion about something—say, the environment or the government. Regions of relations are, however, mere subsets of the endless *expanse* of relations.

---

12. Bradley, *Appearance and Reality*, 27–28. The cognitive scientists and philosophers of mind Eric Margolis and Stephen Laurence wrote, "Either we are involved in a vicious regress, or we might as well stop with the external language and explain its significance directly" (Margolis and Laurence, "Concepts," §1.2). However, to stop simply because we "might as well" is not the soundest of reasons.

13. MacBride, "Relations," §2.

Part II

Personal Ethics

The second part of this book seeks to incorporate— or *reincorporate*—ethics into metaphysics. It does so by surveying an endless expanse of relations. This expanse reveals *meta-features*,[1] which one does not see in a self-contained system.

---

1. As a metaphysical term, "meta" refers to something "more comprehensive" or "transcending" ("Meta," in *Merriam-Webster Dictionary*, 2021).

Chapter 8

An Endless Expanse

An endless expanse of relations reveals various *meta-features*. On the basis of these features, we may derive the broadest of ethical principles.

PHILOSOPHY, BY AND LARGE, has stripped out ethics today.[1] The philosopher David Hume famously wrote that one cannot derive an "ought" from an "is." That is, while we know that this is how the world *is*, how should we know that it is as it ought to be? This is one of the core problems of philosophy: the fact-value distinction—or, more specifically, the is-ought distinction. We call it Hume's law.[2]

Should I take a walk in the woods today, or should I write letters instead? Should I be a "bachelor girl," or should I marry Joe? Should we travel to Mars? Should we drop the bomb? On the surface of it, our reasons for choosing one course of action over another might seem obvious, yet it is not something we find ourselves able to decide on the basis of the facts.

There are more pressing issues. Should we defund the police? Pay reparations? Close the border? Banish nuclear armaments? Reduce our travel? Decolonize education? Shut down autonomous zones? Empty the prisons? Exit the Union?

Since Hume, there has been a growing tendency to downgrade ethics as (ultimately) nonrational—even to banish it from philosophy.

---

1. Bertrand Russell, in his *Problems of Philosophy*, ignored the subject altogether. It did not warrant an entry in the index (Russell, *Problems of Philosophy*, 99).
2. It may also be called the fact-value gap or Hume's guillotine.

Ludwig Wittgenstein wrote that ethics has become "what we cannot speak about." Yet ethics pervades all that we do, from morning till night, from year to year.

This problem has now filtered through to the common person, causing profound disorientation in our time. On a social level, we are conflicted and confused with multiple ethics, while on a global scale, our ethics seems increasingly to have come apart.[3]

This is not to say that, without a rational basis for ethics, we do not apply our reason to ethics at all—however, reason fails us wherever we seek to *ground* our ethics. Our ethics, wrote the theologian Deane Galbraith, is "grounded arbitrarily."

> If we want to argue for a moral position, we
> need to find a rational basis for ethics.
>
> —MEL THOMPSON

Emergence

All major approaches to ethics have a fundamental problem in regard to their rational grounding, or relationship with reason—among them the following:

- *Theological voluntarism* holds that ethics is revealed by God—alternatively, an eternal power.[4] Therefore, moral principles are obtained through *revelation*. But ultimately, revelation excludes reason.

- *Moral intuitionism* proposes that ethics is known through intuition. Yet intuition, or some strong feeling that this is how things *ought* to be, is unsatisfactory from the point of view of reasoned judgment—in fact, may be positively dangerous.[5]

---

3. David Korten provided three major examples of global crisis: deepening poverty, social disintegration, and environmental destruction (Korten, *When Corporations Rule*, 28–32).

4. Variants of theological voluntarism have been presented throughout the history of religion and philosophy. The term "divine command theory" may also be used yet could be misleading.

5. While it may seem obvious to more benevolent souls which ethical propositions are true, it may not be obvious to others.

- *Ethical naturalism* holds that ethics may be inferred from the way the world works, independent of personal opinion. Here, however, we still have the problem of Hume. How do we advance from matters of fact to moral principles?
- *Ethical relativism* holds that ethics is determined by *society*. This view has held increasing sway, as people have, paradoxically, become progressively intolerant of dissent from cultural norms. Again, it is ultimately ungrounded.
- *Utilitarianism* is deeply problematic. I shall call it *strategic rationalism*.[6] Through reason, it seeks the best strategy for the future. This may offer the illusion of simplicity, yet it claims far too much for itself. Designs on the future exceed our powers of thought.[7]

Where, then, to turn?

In earlier chapters, we exploded the limits of words and of linguistic systems. Then, in the last chapter, we sought to assess the *number* of things and the number of relations between them. Now we may step back, to view it all as a whole. It is what we often do, when, proverbially, we cannot see the forest for the trees.[8]

As we step back—that is, as we zoom out—we see the whole of reality and not merely the parts. We see the entire expanse of relations—that is, things and *relations* between things—stretched out before us, all interconnected in every direction. In this whole, then, we discern *features* of the whole—specifically, the broader features of things and relations. We call it *emergence*. The whole reveals properties that the *parts* do not.[9]

An analogy would help.

Imagine that I stand in a field of wheat. I stoop to feel a flag leaf. I grasp a ripened spike in my hand and roll the kernels in my fingers. Then, I lift up my eyes and look around me. I see many more such wheat plants, everywhere.

---

6. This would include utilitarianism.

7. Strategic rationalism would ultimately need to know how to gain control of the whole world, and this is never achieved. Apart from which, there is the problem of choosing the goal.

8. "Always take flight," wrote Wittgenstein, "to where there is a free view over the whole *single* great problem" (Pichler and Säätelä, *Wittgenstein*, 2:108).

9. This definition of emergence is a simple one, and various definitions exist. One will understand the basic idea. The philosopher Nicolai Hartmann described emergence as a *categorial novum* (Poli et al., *Philosophy of Nicolai Hartmann*, 118).

But now, I see something new. I see *rows* of wheat, stretching out into the distance. I see *swathes* of wheat, which are bowed down by the wind, now here, now there—and soft shadows in a depression in the field. I begin to discern features of the *whole*.

In the case of language, and the reality that it describes, we see wider features, which are surprisingly obvious and now give us the conceptual goods we need to develop a new moral philosophy. When we survey the *whole*, we are able to develop, one by one, ten ethical maxims—all of them interlinking and overlapping—which are rationally grounded.

Ethical Maxims

Above all, when we cast our eye over an endless expanse, this differs fundamentally from other ways of looking at the world—in particular, the *bounded* view, which we described in chapter 2, where philosophers interested themselves chiefly in *immediate* things: the nature of matter, the requirements of the "good life," or the principles of politics.

We may add to this all points of view that are limited in *scope*—which cover limited regions of relations—limited, for example, to ethnic, religious, ideological, economic, or scientific points of view.

Once we have stood back, or zoomed out, and apprehended things and relations as a whole, we no longer satisfy ourselves with the narrow view.[10] We see, as it were, all things. No longer is our thinking parochial, short-sighted, or self-interested. Rather, we think about our world expansively and holistically, through all space and time.[11]

At the same time, we now have a view of *words* that describes worlds of relations within—relations that further reach across all the expanse of our language and reality, combining with relations inside other words, everywhere.

This has the necessary consequence that we reject the reductionistic, simplistic, technical thinking of our present time, which reduces words to atoms, cutting away from them a great many relations that they trace—and with that, meanings and significances.

---

10. Immanuel Kant wrote that we seek to extend our knowledge of relations among phenomena (Losee, *Historical Introduction*, 98). The physicist and philosopher Ernst Mach wrote that we formulate relations which summarize the greatest number of facts (Malin, *Nature Loves to Hide*, 5).

11. William James proposed that moral progress is marked by the replacement of less inclusive ideals by more inclusive ones (James, *Will to Believe*, 132).

An Endless Expanse 57

From this perception of an endless expanse of relations, we may now propose the first of ten ethical maxims:

- *Think expansively and holistically.*

Our first ethical maxim leads us naturally to a second.

As we stand back to view relations as a whole, we understand that the very word "relations" speaks, of course, of relations between things. Relations in our world both exist and are ordered in a certain way, as best we can discern.

Yet we do not always perceive it or even accept it—namely, that things are related as they are. Rather than making the faithful ordering of things and relations our goal (which is to pursue truth), we make it people, possessions, culture, or prestige.

We may do two things to *reject* the relations that we find in this world. We may add to them or subtract from them. We can, of course, "cross wire" them, too—that is, make false connections, but this is much the same.

Take, as an example, a simple house, which is built of stones, bricks, and rafters.[12] Now suppose that we claim it has a fireplace, where it does not. We have now added relations—*fabricated* relations between things—which do not exist.

Or suppose that someone sells us tickets to the imaginary planet Vulcan. This indicates an ethical fault—because they have added a planet to the solar system, where there isn't one.

So, too, if someone believes that there is a master race, or that human nature improves with time.[13] There are not the relations to support such views.

Or suppose—returning to the example of the house—that we claim that we have removed its rafters, where we have not, or that it has no foundations, where it does. We then *strip away* relations that exist.[14]

We return to an earlier example. In hotter weather, it thunders in my city. Yet in order not to frighten tourists, I announce that it does not in fact thunder, even on the hottest of days. Since it most certainly does,

---

12. An example of Aristotle's, *Metaphysics*, 216. We list the features of a house differently today.

13. There is the idiom "No rose ever saw a gardener die." There are certain things we cannot tell because of the brevity of our existence.

14. There is another possibility. One may emphasize or de-emphasize relations, inappropriately.

we have stripped away relations that exist—in this case, relations between the heat and the thunder.

We find this everywhere we look. On a personal level, we find economies of truth, misrepresentation, scapegoating, and denial; while on a political level, we find diversion, minimization, cover-ups, book burning, and so much more—where all of these represent the *removal* of relations from the whole.[15]

From this, we may now put forward a second ethical maxim:

- *Embrace those relations that exist in the world.*

Now when we view the entire expanse of relations, we discover that these relations *complement* one another and may be brought into *balance* with one another.[16]

We do it all the time. As I set out on a hike on a cold morning, I do not forget the heat of the noon day. As I beach my boat at low tide, I do not forget that high tide is coming. As I think of the future, I do not forget the past. As I consort with the rich, I do not forget the poor. As I look upon the centralized, I do not forget the marginalized.

It is a natural corollary of our second maxim: *embrace those relations that exist in the world.* As soon as we add relations to the whole or remove them *from* the whole, we not only misrepresent reality, but we affect its *balance*.[17] In fact, we affect our own balance, too.

Our language is replete with reminders—*idiomatic* prompts—that we should beware the loss of balance between things. We may "fly too high," "push the limits," be "thrown off balance," or "crash and burn." And sometimes, too, we "strike a good balance," "give and take," or "meet people halfway."

The value of balance was well recognized by Lao Tzu, who wrote, "When considering any thing, do not lose its opposite."[18] We need "subtle

---

15. One cannot, however, be quite this simplistic about it. These terms may be different things to different people. Nuance is required.

16. This might seem impossible to some, in view of an almost infinite expanse of relations. In ch. 11, we shall consider more closely where one is to begin.

17. One might suggest some extreme possibilities. For instance, "Why not balance the number of free children to slave children?" However, this would leave far too much out of the balance: countless considerations individual, social, political, scientific, and more.

18. Laozi, *Tao te Ching*, §28.

analyses," wrote the philosophy professor Daniel Graham, "revealing the interconnectedness of contrary states in life and in the world."[19]

A sound ethics is lost where *balance* is lost—in fact, every unvirtuous or evil deed may be traced back to a loss of balance—or to put it another way, to either the exclusion, or the unfounded addition, of relations between things.

From this, we may now advance a third ethical maxim:

- *Seek to balance all things.*

It may not, however, be quite as simple as it seems. When we omit things from our thinking, we may not even know it or suspect it.

A major example is our pursuit, for centuries, of activities that have compromised or destroyed the environment—relational blind spots through which we have failed to see the relationship of our activities to our surroundings.

This applies equally to problems of social disintegration and deepening poverty, wherever we find them. In fact, it applies to every trouble under the sun. There were connections we did not see.

The philosopher and physicist Karen Barad highlighted the fact that some things are visible to us, others are hard to see, while yet others we may not see at all.[20] To put it more pointedly, it is often I myself who do not see. I imagine that I am informed, enlightened, progressive, critical, while in fact I am visually impaired.

I consider subject *a*. I could, of course, consider not-*a*. High tide? Low tide. The future? The past. The rich? The poor. The prioritized? The marginalized. But this is the easy part.

We should not think that this is the beginning and the end of it. There are other thoughts to think. High tide? It's higher now than it was. The past? It's not what we think it was. The rich? It is not only riches that describe them. The marginalized? We might not so much as have heard of them.

We give closer attention in due course to relations that we routinely overlook. In the interim, we may propose a fourth ethical maxim:

- *Find relations that lie beyond the obvious.*

---

19. Graham, "Heraclitus," §3.2. This does not mean that we admit *contradictions* in our scheme. Rather, Heraclitus held that opposites coexist in what is one and the same.

20. Barad, *Meeting the Universe Halfway*, 87.

But set aside, for a moment, the way in which we view the world *around* us, and let us again look *inward*.

We need to have a good grasp of those relations that exist within our own selves—which is to say, we need to know what we ourselves have to do with the relations that we trace in this world.

For instance, I can ill afford to begin my quest for truth—whatever this quest might become—with an overestimation or underestimation of self. In any area of life, I need to have a reasonable assessment of *myself*.

Too many people have rushed down a road ill prepared, while others have dithered or doubted where they could have seized the moment. Some have been blind to their ignorance, while others have been distracted by fruitless detail. The examples will be as many as the people themselves.

I need to have a sound estimation of who I am, of my powers and limitations, and of my place in the world. If I do not, there are many errors of judgment that await. "Know thyself," proclaimed the Delphic maxim. Progress is not founded on illusions. It must begin with a face-to-face encounter with reality and a fair assessment of self.

True life begins not with *imaginings* of who I am; it begins, wrote the hymn writer Charlotte Elliott, "just as I am, poor, wretched, blind"[21]—a self-assessment that has resonated with millions of people.

From this, we may advance a fifth ethical maxim:

- *Know how you yourself are situated in the world.*

---

21. Elliott, *Selections*, 1.

Excursus

Thus we have five ethical maxims in hand, in which we seek to bring all things into relation with all things. But is it quite this simple? Might we not have decided up front which things are worth relating and which not? Does it not come down to a kind of moral intuitionism, after all? The answer to this is threefold.

In the first place, this is not an ethics that is based on what it is that matters—which is to say, has value—by which we strive (for example) for wisdom, love, duty, or self-fulfilment. Rather, it is based on things and relations, and their meta-features. This is far removed from human values.[22]

Secondly, we shall see in the coming chapters that, where my behavior changes, it is on the basis of the relations that I trace in my mind. In particular, my behavior changes where I broaden the relations that I trace. This means that, by the very fact of "seeing" an endless expanse of relations, my ethics changes.[23]

Thirdly, we shall see in chapter 36 that our reason exists for the purpose of bridging or filling gaps. This includes our ethics. Briefly, our actions are determined by the gaps that we find—that is, the situations and surroundings we encounter. In this, we do not ask, "Ought my situation or surroundings to be different?" But we are getting ahead of ourselves.

---

22. It is as far removed as the natural sciences are from ethics. In the natural sciences, we do not generally ask "Ought I to study these relationships?" or "Should I push this experiment further?" It is about knowledge rather than judgment.

23. Therefore, this book itself may change one's ethics, with no futher reflection required.

Chapter 9

Human Limitations

The finitude of our own thoughts contrasts sharply with a vast expanse of relations. A failure to understand our limitations holds great danger.

THE NUMBER OF THINGS in this world, and the almost infinite relations between them, now serves us in a different kind of way. It is clear that we have neither the power to understand nor the means to control boundless relations in an endless expanse. In fact, not with all the computing power in the world—or all the machines.[1]

We make a great mistake if we imagine that we have powers that we do not. It leads us not only into error but moral madness.

The simple yet crucial idea of *number* now has important consequences for the development of what we shall call a negative ethics. While a positive ethics tells us what we *ought* to do, a negative ethics tells us what we ought *not* to do. Five further maxims follow.

As we zoom out from a boundless expanse, to take the bird's-eye view, we observe that relations—and the things that are related—are almost infinite, yet we ourselves are finite. Therefore, there is a *mismatch* between the number of relations we find in this world and the number of relations we can trace in our minds. It is certainly impossible to trace *boundless* relations in our mind. To achieve this, we would need a mind (or a computer) greater than the universe itself.

---

1. The emphasis is on *control*. While we ought to think about our world expansively and holistically, this does not imply that we know enough in all situations to act.

Human Limitations

Our tracing of relations between things is always incomplete and partial. At best, we are able to trace only modest *regions* of relations in the endless expanse.

In fact, these regions of relations themselves may be incomplete and partial. Like holes in a map, there are gaps in our knowledge and experience.[2] We fill these gaps with hopeful guesses—*believing* guesses—without which life would come to a halt.[3] "I must have left my keys in the garage," I say to myself, without really knowing where they are; or "The rains will come this winter," as I sow my seed; or (a pre-scientific guess), "I am able to see, because my eyes emit invisible beams."[4]

This filling of the gaps—if we should even notice that we do it—we call intuition, on the one hand, or conjecture, on the other. Our intuitions and conjectures will, of course, not always be true—as was the case with eye beams, which we now know not to exist.

There will always be relations that lie beyond our power to explain and beyond our control—and relations that we fail to take into account at all. Therefore, our tracing of relations will create unintended consequences, whether personal, social, political, or environmental.[5]

Nobody foresaw smog, when the internal combustion engine was invented; or radioactive contamination, when we first split the atom; or ozone depletion, when we first used chlorofluorocarbons (CFCs) as refrigerants.

In fact, unintended consequences have reached into every personal, social, and environmental sphere. In many areas, our human actions have had unforeseen and unpleasant side effects, sometimes only many years later—while tomorrow's unintended consequences, which as yet we do not see, are already being created today.[6]

---

2. The philosopher-physician Francisco Sánchez held that we understand nothing completely (Sánchez, *Nothing Is Known*, 238).

3. The philosopher Arcesilaus held that, even in the absence of genuine knowledge, one can think and act (Brittain and Osorio, "Arcesilaus," §5). While our grounds may not be certain, they may be reasonable.

4. Many people still believe this. It is the emission or extramission theory of vision. Ibn al-Haytham, in the early eleventh century, was the first to propound a view that ultimately overturned this (Lindberg, *Beginnings of Western Science*, 313–14).

5. Stephen Hawking noted, "Even if we do achieve a complete unified theory, we shall not be able to make detailed predictions in any but the simplest situations" (Sobosan, *Romancing the Universe*, 175).

6. At the time of writing, the dangers of SF6 were revealed—a potent greenhouse gas used as a dielectric medium with "clean" electrical gear. It came as a shock for many advocates of clean energy (Dervos and Vassiliou, "Sulfur Hexafluoride," 137).

This has important consequences for ethics. A sound ethics will recognize the danger of our limitations—such that our actions will be guided by a deep sense of inadequacy and the desire to know more.

From this, we may derive a sixth ethical maxim:

- *Be guided by a deep sense of inadequacy.*

Not only do we need to be mindful of our *lack* of control; there is, further, the danger of *excessive* control. Here, too, our appreciation of an endless expanse of relations gives us direction. It reminds us that we should be wary of our totalizing urges, which seek—impossibly—to conquer boundless relations. In order to *conquer* boundless relations—to have *mastery* over them—we would need boundless control.

This is impossible. The philosopher Epictetus wrote, "Happiness and freedom begin with a clear understanding of one principle: some things are within your control, and some things are not."[7]

Where we strive for an unreachable mastery, unintended consequences disrupt our ambitions and set the bar for control even higher—unnaturally, impossibly higher. As a result, our totalizing tendencies always have to do with damaged lives and disasters, as we push against the bounds of the possible.

In its more innocent guise, we may find this in our ideas of progress, advancement, development, even utopia[8]—in which we underestimate the danger of our limitations and open the door to hubris and failure.

More obviously, our world is full of symptoms of our drive towards absolute mastery and systematic totality: among other things, the totalitarian state, excessive legislation, ruthless censure, extensive agricultural controls, industrial espionage, weapons in space, aggressive competition, and, of course, control freaks—all by-products, in some way, of the inability of finite creatures to master the infinite.[9]

We all have personal experience of the same. The card collector must have every card in her collection, the cook must empty every last

---

7. Paraphrase of Epictetus, *Art of Living*, 3.

8. This includes social optimists such as Hegel, Comte, Marx, Peirce, and Teilhard de Chardin.

9. The philosopher and psychiatrist Karl Jaspers wrote, "The overweening plans of rulers based upon a supposed total knowledge of history have always ended in catastrophe. The plans devised by individuals in their restricted circles fail or else contribute to unleashing quite different unplanned complexes of events" (Jaspers, *Way to Wisdom*, 104).

bean from the can, the golfer must repeat his swing until he reaches consummate perfection, while the bride will not enter the church until she has every wisp of hair in perfect place.

From this, we may propose a seventh ethical maxim:

- *Avoid totalizing tendencies.*

The dangers that lie in our lack of control and the temptation to excessive control both lie in *overreach*. In our limited capacities, we assume that we have mastery over areas we do not. In fact, there are many things in life—*most* things—of which we know nothing at all.

Perhaps my neighbor understands the city's water supply network. I have no knowledge of that. Another neighbor might know how many tons of wheat were harvested north of the city this summer. I had not thought of that at all. Another might know that all the surgeons left the local hospital. I know nothing of that either.

There are many things that lie beyond my *ability* to include them in the scheme of things. Or, to put it another way, the way in which I trace relations between things will have to be, in an important sense, *subjective*. I cannot survey all things at once, to bring about the most perfect integration of them all. I can begin only where I am.

So it should be. Although I live in a world of boundless relations, it is not possible to make the entire field of relations my focus. Therefore, there is always one thing with which I must begin, when I trace relations between things. We call it a *relatum*. We need one more relatum, then, to bring two relata into relation with one another.

Where do I begin, as I seek to understand my world? With few exceptions, I can begin only with relata that I already have in hand.

This provides us with an eighth ethical maxim:

- *Start where you are.*

Avoiding Centers

There is another reason, apart from our unknowing, why we must start where we are. This lies in a more academic problem, as it were.

As we survey an endless expanse of relations, we find no reason to plant our flag in one place rather than another. We see no reason to choose this landmark or that, this center, that origin, that starting point. To put it simply, we find no place to begin, *in principle*.

Apart from this, where there is an endless expanse of relations, it is impossible to see its end. We cannot see *far* enough to know where to plant our flag.[10]

An endless expanse of relations is unable, therefore, to provide us with a stable center: the sun, say, the king, my ambitions, or my lover.

Countless people's lives have revolved around such centers, and all such centers have come and gone. All have failed us and disillusioned us on the way: sacrifice to the sun, massacres in the name of the king, shattered ambitions, or a broken heart.

> [Metaphysics is] the enterprise of returning "strategically," "ideally," to an origin or to a priority thought to be simple, intact, normal, pure, standard, self-identical.
>
> —JACQUES DERRIDA

Our very *language* reveals that there are no landmarks, no centers, no origins, no starting points. Our very language is, so to speak, *flat*. It is a *conceptual* flatness, where no concept reduced to words finds any reason to rise above another.

Our words do not carry special marks to distinguish them in that way. We find no *labels* on them, such as "primarius" or "secundus," to indicate their importance. We see nothing there that carries special weight. This again sets our language apart from the language of the philosophers who privileged certain concepts in their philosophies.

It stands to reason. The very things that words contain have no priority over one other. They do not come with marks that indicate their greater importance or higher priority. They are all of them composed of mere matter, after all.

This has one major consequence for ethics. In a world with no centers—with no special foci or any compelling reason to create them—there are many things that carry no intrinsic value: not social status, race, technological progress, financial leverage, national pride—in fact,

---

10. We have noted the problem of Hume's law, too. As soon as we attribute to something a greater weight than anything else, we break this law.

not my very life. All such, where we unduly elevate their importance, are false values.[11]

In fact, the more that I elevate any such value in the endless expanse, the greater may be my fear that it is empty and illusory. And where I fear that my feet may be standing on nothing, there, the risk of *fundamentalism* arises[12]—where fundamentalism is our resistance to the fear that there may be nothing in it.

And then, our stable centers continually have the habit of displacing *other* such centers. We saw this in the case of the philosophers who continually put each other down. They could not, so to speak, coexist, even if they were separated by centuries.

This has been well recognized by postmodernism. My own centers suppress, exclude, and marginalize others. For instance, business avarice will tend to exclude environmental interests, political dogma will tend to repress human rights, personal ambition will tend to marginalize my own closest friends, and so on. In the words of philosopher Max Black, "To see anything well is necessarily to ignore what you don't want to notice."[13]

This is especially pronounced in *theology*. The language of personal salvation may suppress the language of social commitment, the language of community may exclude the language of justification by faith, or the language of religious values may marginalize the language of the glory of God. Theological language has a natural, powerful tendency to exclude other theological concepts.

Therefore, an ethics of relations is an ethics of *equality*—in the sense that no stable center may take priority or pride of place in my world.

From this, we may put forward a ninth ethical maxim:

- *Avoid the creation of centers, origins, or landmarks.*

---

11. *In principle*, we find no nodal points or foci in an endless expanse of relations. Later, we consider why this is not so in practice.

12. *Religious* fundamentalism comes most easily to mind; yet we may think of ethics, too. Where ethics is not rationally grounded, our actions are at risk of being exposed as empty. Thus we may defend our ethics with censorship, coercion, or violence. We may all be vulnerable to this fear.

13. Black, *Prevalence of Humbug*, 87.

Excursus

Yet are not "relations between things" a central idea? Is not everything in this book referenced to relations between things? There are three things that distinguish things and relations from our familiar centers, origins, or landmarks.

Firstly, things and relations contain all central ideas, yet central ideas do not contain all things and relations. Things and relations are common to all philosophies and include their central ideas—yet not all philosophies survey the entire expanse of relations.

Secondly, many philosophies work within a fixed frame of reference, which is *structural linguistics* by its broadest definition. A philosophy of relations *exits* this fixed frame, to survey the almost infinite expanse of relations from the bird's-eye view. It starts not with the single thought, but with the "everything" of the title of this book.

Thirdly, we shall see—in part 5 of this book—that we must *dissolve* both things and relations, in order to solve some core problems of philosophy. There will be nothing left of things and relations in the end. We shall think on this more carefully in due course.

Shifting Relations

Lastly, as we survey an endless expanse of relations, we see that a vast number of relations is not fixed but is *changing*. In chapter 7, we noted that there are new things—new concepts—continually coming into existence.[14] This is a world that is always shifting and changing—always appearing in some new way, under our very eyes.

While some relations will presumably never change—a molecule always has the same bond distance, for instance, and life will always die without water—social dynamics, modern technology, popular opinion, the familiar routines of the everyday, even the positions of the stars do not remain the same.

---

14. And we may add, slipping out of it again, in many cases.

Any philosophy that is based on relations must take into account the shifting nature of reality. Today's philosophy may not address tomorrow's needs, and yesterday's philosophy may not address today's.[15]

The discovery of the relation of energy to matter ultimately gave rise to many ethical problems—among them, the possibility of atomic war and radioactive contamination. New technologies to control conception and gestation raised high-profile moral issues surrounding contraception, surrogacy, and more. New possibilities for surveillance brought increasing social dilemmas—among them social engineering and the exploitation of big data. The list is vast and growing.

We would do well, too, to look *inward*. If the world is always appearing in some new way, we should reexamine the way that we cope with it and accommodate it. If we continue to approach the world in the same way, when the world is not the same, we may do harm both to ourselves and to others.[16]

From this, we may derive a tenth ethical maxim:

- *Avoid fixity.*

We now have ten ethical principles in hand—which are rooted in the broader features of relations as a whole—that is, relations between things:

- *Think expansively and holistically.*
- *Embrace those relations that exist in the world.*
- *Seek to balance all things.*
- *Find relations that lie beyond the obvious.*
- *Know how you yourself are situated in the world.*
- *Be guided by a deep sense of inadequacy.*
- *Avoid totalizing tendencies.*
- *Start where you are.*
- *Avoid the creation of centers, origins, or landmarks.*

15. Thomas Kuhn noted of scientists who occupy different paradigms, "Both are looking at the world, and what they look at has not changed. But in some areas they see different things, and they see them in different relations one to the other" (Kuhn, *Structure of Scientific Revolutions*, 150).

16. William James held that truth evolves—insofar as new facts clash with old beliefs and so require readjustment (Putnam, "James, William"). Rabindranath Tagore wrote, "Life, moral or physical, is not a completed fact, but a continual process" (Tagore, *Complete Works*, 1202).

- *Avoid fixity.*

We close with a counter-analogy to that with which we began.

Suppose that I am a virologist, with a *specialized* interest in wheat. I stand in a field of wheat. I stoop to feel a flag leaf. As I stroke my thumb over its surface, I feel a fine ripple in the leaf.

I pluck the leaf and hold it up to the sun. I see streaks of yellow and green—telltale signs of contagion. I seal the sample in a small container, drop it into my pocket, and return to the lab.

As a virologist, I have little interest in the broader features of the field—whether its rows stretch out into the distance, how it is bowed down by the wind, or whether the sun casts soft shadows in a hollow. I have the special interests of a virologist. I relate things to things within a strictly defined scope.

Applied to philosophy, it means that we are unlikely to see the great expanse of *all* relations when we are confined to a limited—or limiting—field of view. We are unlikely to discover *emergence*. This is precisely what philosophy has done in the past. It has joined things to things, without taking the bird's-eye view.

Chapter 10

Relations and Expectations

Whenever we hold up our conceptual arrangement of the world to the world itself and there discover what we do not expect, we are moved to act.

THERE COMES A POINT where ethics meets reality. Ethics finds its outworking in the world. What is it, then, that *connects* our ethics to our actions? What is it that *drives* us? What is it—whether intrinsic or extrinsic—that motivates us to act? What is it that motivates me to plant a garden (and to plant it *thus*), to embark on a career, or to go to war? What is it that, as it were, joins the engine to the wheels?

I could easily be confused. I am motivated by the smell of cookies, a love for humankind, greater rewards, the fear of death, revenge on my enemies, and countless things besides.

Yet our motivation is ultimately based on something far simpler than this. It is as simple as the law of noncontradiction—or perhaps, rather, the principle of contradiction. Where two things do not match in our minds—specifically, where the world as we perceive it does not match our expectations—there, we are motivated to act. To put it simply, our behavior is controlled by *mental models*.[1]

The simple principle of contradiction is vital to our understanding of our actions. Without it, we may fail to associate human behavior with

---

1. Strictly, if things are concepts and concepts are things, as described in ch. 3, then the real world with which I am at odds exists in my mind.

our outlook on the world. Thus we may fail to understand the *reasons* for people's behavior—and fail to address it appropriately.

Philosophically, as to what motivates us, there have been two major points of view. David Hume believed that we are incapable of following a rational code of ethics. "Reason," he wrote, "is and ought only to be the slave of the passions."[2] The philosopher Immanuel Kant, on the other hand, considered that we are moved by an impersonal, exterior source of normative principles—which is reason.[3]

Many have tended towards Hume and many towards Kant. Which, then, shall it be?

The influential psychologist Richard Lazarus and many others since have held that our visceral (gut) feelings are generated by novelty, discrepancy, and interruption.[4] In each of these cases, we are dealing with relations between things that we did not *anticipate*. That is, if we find that things are not so arranged as we think they *ought* to be, we react viscerally.[5]

In fact, even a dog, when faced with food it does not expect to see in its bowl, is visibly affected. All in all, we may summarize it as encountering the *unexpected*—and if not the unexpected, then, as the philosophers Willard Van Orman Quine and Joseph S. Ullian put it, "something expected [that] fails to happen."[6]

I enter my home and am surprised to see a new painting on the wall: this is novelty. I arrive at a board meeting and am shocked to see the police there instead: this is discrepancy. Or I take a flight to another city, but it is diverted: this is interruption. In short, it is the relations that we anticipate in this world, when we hold them up against the world, that have everything to do with emotion and motivation.

Once, then, the *motivation* is there, I look for opportunity to *realize* it—and always, my *body* is instrumental in carrying this out.

2. Hume, *Enquiry*, 3. Hume may have been paraphrasing the philosopher and mathematician Blaise Pascal, who wrote that one's senses are "never the servants . . . of reason" (Pascal, *Thoughts*, 186).

3. Philosophers Robert N. Johnson and Adam Cureton summed up Kant, "Only reason itself has genuine authority over us" (Johnson and Cureton, "Kant's Moral Philosophy," §14).

4. J. Harris, *Developmental Neuropsychiatry*, 1:122.

5. However, we may not feel as strongly about one thing as we do about another. In this case, we might say that we *prefer* that things should be arranged like this or like that.

6. Quine and Ullian, *Web of Belief*, 20.

Relations and Expectations

I decide to move a table. I may push it to the corner of the room, or I may ask someone to do it *for* me (speech is a bodily function, too). More recently, I may need merely to generate *cerebral* activity of a certain kind, to make things happen.

Of course, I may not always be able to realize my intentions. Someone may have bolted the table to the floor, say, or I may suffer a dizzy spell that prevents me from moving it. In such cases, I may have to revise my actions—or revise the way in which I arrange the world in my mind: "Actually, I always thought that the table looked better where it is." Thus we conform ourselves to that which we find in front of us.[7]

We may be able now to resolve the age-old conflict between reason and passion.

When we hold up our conceptual arrangement of the world to the world itself and find a difference there, we are motivated to act. A conceptual arrangement of the world thus modifies our motivations. In this way, reason and passion are both master and slave. While reason does not directly control our passions, we may trace our passions back to reason.[8] Simply put: reason in, passion out.

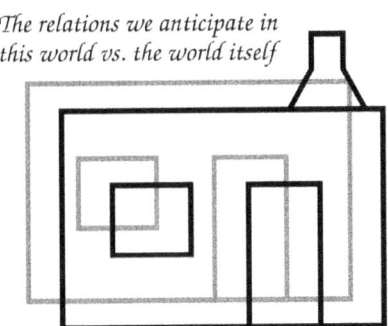

The relations we anticipate in this world vs. the world itself

The Problem of Others

It would seem, then, a fairly simple matter to understand our motivations—in principle, at least. We find that things are arranged thus and thus—and depending how we think they *ought* to be arranged, we may

---

7. Or, as Ovid wrote, "*Video meliora proboque, deteriora sequor.*" We may not follow the course which we approve, because of the weakness of our will (Shaw, *New Dictionary of Quotations*, 465).

8. To put this another way, reason does not give us power over first order desires. In an important sense, behavior is not born at the level of conscious thought.

understand how and why we act. That is, wherever we find a discrepancy between reality and expectation, we typically have some kind of visceral reaction, and we act.[9]

What, then, shall we make of more *spontaneous* actions? That is, are there not actions that are more impulsive and instinctive than our identification of discrepancy? What of the case where someone treads on my toes, and I push them away—or I feel hunger and grab a sandwich?

While many situations would seem to be more instinctive, they, too, will often represent a discrepancy between reality and expectation. If an army instructor intends me to bear pain in my toes, my expectation is different—and so is my behavior. If a day of fasting requires that I should leave sandwiches alone, I do not go grabbing a sandwich when I feel hunger. Even seemingly spontaneous actions, therefore, often have to do with what we *expect*—alternatively, do not.

However, our focus has been fairly narrow, until now. Motivation is more complicated than this. Above all, mine is not the only mental model in this world. There is the complication of *others*—that is, other *people* in this world.

We return to the table in the room. Consider now that there is another person *with* me in the room, who thinks that I ought *not* to move the table to the corner of the room but that it ought to stay in the middle. I now need to take into account the will, desires, and beliefs of another person. Let us suppose, too, that this other person feels so strongly about it that they *sit* on the table to prevent me from moving it. This will have a profound influence on my own motivation to move it.

As I survey my world, therefore, not only do I see things arranged around me, but I must deal with other relation-tracing beings. While I may sometimes see others as mere objects among objects, I need mostly to take into account the complex and subtle ways in which they, too, are involved in ordering my world.

Imagine a game of fast chess, where a timer is used. A grand master surveys the board and reaches out his hand to make his next move. But in this very moment, just before time-out, we permit a novice to insert an additional move from the side. The grand master's challenge is now vastly greater. He did not anticipate the novice's move—nor is he accustomed to the increased demands on his faculties.

---

9. Many have ascribed intentionality to emotions. Philosophy professor Rami A. el Ali called it "emotional intentionality" (El Ali, *Intentionality of Emotions*).

So, too, is the nature of motivation. *People* are continually modifying our game.

We have just made an important distinction. On the one hand, we arrange the world in our mind. On the other, others arrange their worlds in *their* minds. These others may be one, two, or several or may include the whole world. Yet no matter how few or how many, their acting in this world introduces a quantum leap of complexity as we seek to understand our own world.[10]

Theoretical physicist Stephen Hawking wrote, "While physics and mathematics may tell us how the universe began, they are not much use in predicting human behaviour, because there are far too many equations to solve."[11] Such equations would require not only that we observe people's outward behavior but that we take into account their inner worlds.

And then, not only do others seek to arrange the world in their minds, but they have strange ways of *communicating* what they are doing. While we say that nature is a book, which (ideally) may be read, not so with people. Much of the way in which they have arranged their world in their minds needs to be inferred from what we call *semiotic codes*: gestures, postures, commodity codes, esthetic codes, and so on—as well as language itself.

A frown, a jig, a nod of the head—Chairman Mao's swim in the Yangtze river, the bomb under Mururoa, Nelson Mandela's tea with Betsie Verwoerd, and a host of interpretive devices all make further demands on our understanding of the world.

What drives me, therefore, is not only the way that I trace relations in the world around me, not only the way that other people do, but the *semiotic codes* through which others reveal their worlds to me.

This has an important consequence. All the science of motivation and manipulation cannot replace human rapport, responsiveness, and reciprocation. The subtleties of personal interactions now move closer to center stage.

---

10. The historian and psychologist Wilhelm Dilthey emphasized the radical distinction between the natural and the human sciences. "Regarding all other objects there is an interest to explain; regarding human beings, an interest to understand" (Dilthey, "Hermeneutics," 4:229).

11. Hawking, "Stephen Hawking Quotes," quote 2.

Inheritance

It might have slipped our attention. So far, our focus has been on that which motivates us *in the present*—or, rather, we have omitted any thought for the past or the future. Positioning a table in a room, playing a game of chess, and so on, are all examples in the *present*.

Now words, as we have described them, surely must include the past and the future. Words, we have said, have an inner life, which we might never suspect by merely glancing at their outer form. Words contain a world of relations inside. We trace relations, we said, not merely through the *sentences* we create, but words themselves contain countless relations.

Now consider that, by far, most of the words we use existed well before we knew them—as did their relations inside. They previously existed as worlds of relations inside other people's minds, the minds of people who may for the most part no longer be with us. This means that many of the relations that I trace—that is, the words that I speak—reach me *ready-made*.

> There is a certain memory in a culture
> that is carried on in its language.
>
> —FRANCIS SCHAEFFER

I may add these words to my vocabulary without any special awareness on my part. I learn what a house is or an apple, a dog, or a moon—not to speak of a host of abstracta—with a sophistication far beyond anything in the dictionary, without (mostly) giving it a thought.

In an important sense, therefore, many of my words—which are the relations that I trace—are not my own but are inherited from my forebears.[12] This goes back a very long way. Both the form and meaning of a word may still be recognizable after thousands of years. In the case

---

12. Compare with Plato's belief that all general truths represent the recollection of an existence before birth (Blackburn, "Anamnesis," in *ODP*, 15). Johann Herder held more generally that a language expresses the culture and character of a nation (Coleman, "Disneyland of Cultural Rights," 57). Alexander von Humboldt held that a language embodies a distinctive worldview (Kates, *New Philosophy of Discourse*, 165).

of English, many words still bear some resemblance to Sanskrit. Take, for example, the words *gow*: cow, *lubh*: love, *mithya*: myth, or *danta*: dental.

Now because words are not atomic—because they contain a world of relations inside—including those relations that were traced in the past—we inherit the *thinking* of previous generations.[13] To put it another way, *culture* is transmitted through words.

Thus we find that a matter as simple as the nature of the words that we use explains the transmission of culture and all that culture brings with it: industry, sociability, isolation, even migrations and wars.[14]

This is not to say that we inherit *all* the relations that a word may contain nor that we do so all at once. It typically takes *time* to learn words in all their fullness—and not all of a word's relations will be inherited. Yet because words are *more or less* inherited, they may contain the wisdom of the ages, which enables us to survive and to thrive—or they may contain outworn ideas and folly, driving us to intellectual decline, the destruction of the environment, or the subversion of familial relations, among other things.

This would suggest some affinity with the doctrine of original sin, which holds that human nature is morally corrupted from the start. In much the same way, we inherit ways of thinking and doing that do harm to ourselves, our society, and our environment—even to the point of evil.[15]

Can we therefore shake off our outworn meanings and patterns of thought—our arrangements of things in our minds—if they enter our language through stealth from the past?

We can indeed—on condition that we are willing to renounce every relation we have ever traced. This is the first step of philosophy. Philosophy is not about mere addenda to our thoughts or adjustments to them. Philosophy should disorganize and reorganize our fundamental orientations for living in this world. The philosopher, wrote the

---

13. It seems worth considering that this may be the source of a belief in reincarnation.

14. A necessary corollary is that, if our culture drives us to obliterate an enemy, the enemy endures within our own soul, so to speak.

15. Paul Feyerabend wrote about a "concealed essence" that transcends historical periods (Feyerabend, *Killing Time*, 49). It seems possible, therefore, that societies that experience revolution, whether peaceful or violent, may find that major features of the old society survive revolution, in this concealed essence. This may also explain an apparent fixation with pre-revolutionary dispensations.

philologist and philosopher Friedrich Nietzsche, is a terrible explosive that endangers everything.[16]

We must be willing to abandon and *continue* to abandon culture, wealth, prestige, comfort, even "truth" so-called—in order to arrive at truth.[17] If we value anything more highly, it will tend to restrict, confine, reduce, and distort the truth, through our own prejudices and preconceptions. When we truly seek the truth, our whole being is at stake. Philosophy is not a mere pastime.

---

16. Socrates, less dramatically, likened himself to a gadfly, which provokes a horse by its sharp little bites: "I am the gadfly of the Athenian people, given to them by God" (Plato, "Disinterestedness of Socrates," in *Dialogues of Plato*, 2:124).

17. I define truth, in the glossary, as "the final destination of philosophy, whatever this would be."

Chapter 11

Opaque Relations

It is one thing to have an abstract ethics in hand, quite another to see how it applies where relations between things seem opaque. The devil is in the details.

AT THIS POINT, THERE is a simple assumption that could derail us and prevent all further progress towards a complete metaphysics. This is the common notion that we do not understand—indeed, *cannot* understand—why we make certain choices, let alone have certain preferences.

We have spoken, so far, about ethical principles that are not difficult to put into words. In reality, however, a large part of our ethical thinking is subconscious or unawares, and often enough, it would seem to be quite opaque. Sometimes, no amount of thinking seems to reveal on what basis we trace relations between things.

Should I join the resistance, or should I care for my aged mother instead?[1] Should I throw a life jacket to the parson or the doctor? Should I shop at the greengrocer's or at the supermarket?[2] Is there any way that such things may be decided?

---

1. A familiar example of Jean-Paul Sartre (Sartre, *Basic Writings*, 16).
2. At the same time, particular situations may contain various valid options. William David Ross likened it to striking a ball in a cricket game. There are various ways to strike it without being wrong (Mautner, "Ross, William David," in *PDP*, 490).

The philosopher Jean-Paul Sartre put it like this: "If values are uncertain, if they are still too abstract to determine the particular, nothing remains but to trust in our instincts."[3]

If this should be the case—namely, that our values are too uncertain or too abstract to be of much use to us—then we may not be able to integrate all things under one general scheme of thought. Not only this, but as we surrender our reason to the unknown, we may be dangerously exposed to *irrational* decisions.

The Real Start

Given my *particular* situation, where do I begin?

Where do I begin to set my focus today? What is my best policy going forward? Or, more philosophically, which relations in this vast expanse are worthy of my attentions now? What may I justifiably put out of my mind? Famine in a far land? The ecological crisis? A lover spurned? And what if I am just mightily confused?

Whatever the situation in which I find myself, it is possible to find some basic direction by turning to our first ethical maxim:

- *Think expansively and holistically.*

This may seem vague—yet it *need* not be. It is possible to pursue this maxim both rationally and systematically. We offered a clue as to how in chapter 5, where we noted that, if we discard the predicates of a subject too soon or cast away their opposites, we may overlook important aspects of what we have in front of us—whether it be a power plant (Chernobyl), a ship (the *Titanic*), or a head of state (Putin).

Therefore, whenever we deal with predicates—such as "pleasant" ("the weather is pleasant"), "powerful" ("the patriarch was powerful"), "heavy" ("this crate is heavy"), and so on, we may relate them to their semantic *opposites*—or rather, *oppositions*[4]—to help us broaden our view.

These oppositions are clear and specific, so that we may proceed fairly scientifically, if we please.

---

3. This echoes Hermann Hesse: "Knowledge can be communicated, but not wisdom" (H. Hesse, *Siddharta*, 107). At least, wisdom requires lengthy maturation.

4. The word "opposites" may be misleading. Professor of linguistics Sebastian Löbner preferred "oppositions" (Löbner, *Understanding Semantics*, 87–93).

> ### Excursus
>
> We find oppositions of various *types*: antonyms, directional opposites, complementaries, heteronyms, and converses. Imagine, for example, that we read that a certain politician is "influential." The predicate is "is influential."
>
> - *Antonyms*: the politician may be weak-willed instead or may not bring about much change.
> - A *directional opposite*: influence aside, he may be *impressionable*, too.
> - *Heteronyms*:[5] others—even circumstances or divine will—may be influencing his situation, too.
>
> Having now identified some oppositions to "influential," we may ask whether any apply to the politician. This might reveal that he did not bring about the change that we *thought* he did or that he faced strong opposition. We now begin to think more expansively and holistically—or, one might say, with greater nuance.

The same applies to the *subject*. As we focus on a subject, we do well to consider its oppositions and bring them into the picture.[6] For example, when we speak of war, we may think of peace (an antonym); when we speak of advance, we may think of retreat (a directional opposite); or when we speak of Sunday, we may think of Monday or any other day of the week (heteronyms).

Therefore, wherever we are, in whatever situation, we no longer satisfy ourselves with the narrow view. We *broaden our horizons*—and through the use of oppositions, we always have a place to start. The scientist may pay closer attention to the arts, an ideologue may look more closely at human experience, or a captor may be touched by a captive's pleas—to give but a few examples.

---

5. Heteronyms may be variously defined. Löbner defines them as "members of a set" (Löbner, *Understanding Semantics*, 91–92).

6. The social theorist and philosopher Isaiah Berlin emphasized that ultimate values are incompatible (Berlin, "Berlin, Sir Isaiah"). Therefore, one needs to balance opposites—such as liberty and equality, justice and mercy.

Thankfully, we need not depend entirely on our own insights. There were people before us, who traced uncountable relations between things, from which we ourselves may benefit: among others, scientists, sociologists, artisans, and technicians. Therefore, as I seek to survey a wider expanse of relations, I look to what others have learnt. What have *they* said about this? How may *their* insights correct my thinking or expand it?

The Subconscious

Set aside, for a moment, what we *do not* know, and consider what we *do*. We know that our world represents an endless expanse of relations. We know, too, that we base our ethics on the way that we trace relations in this world. Here we have a low-level concept—a philosophical basic—that is the description of our world in terms of things and relations.

We know that we act as we do, because we trace relations in a certain way. Therefore, while we may not understand the relations that we trace, we know that they are present in all our choices.

Suppose that I choose to shop at the greengrocer's rather than the supermarket. While I may not know just why I do it, I do know that I do it on the basis of the way I have arranged the world in my mind—that is, the way I have traced relations between things. There is something about the arrangement of the world in my mind that leads me to shop at the greengrocer's. This may not lead me out of all unknowing, but it may set me on a helpful path—namely, to find the relations that I know I have already traced.

While the simple assumption is superficially true—that we do not understand, indeed *cannot* understand, why we make certain choices or have certain preferences[7]—*in principle*, we know. *In principle*, the answer lies in the way in which we trace relations in this world. This enables us to incorporate all of our behavior in a single scheme, even if we do not explain it in detail.

---

7. Another way of putting this is that we have the concepts but do not have the words or labels for them. Godfrey Vesey and Paul Foulkes considered, "People can use [concepts] correctly though they cannot define them" (Vesey and Foulkes, "Concept," in *COLDP*, 65).

> Concepts can be linked by other means than logical relations of entailment ... and while such relations are non-logical, they are, of course, not necessarily illogical.
>
> —ANDREW SAYER

There is, apart from this, another way I may be helped. The answer to some of our more clouded thinking lies in the *function* of things. We naturally arrange our world to fit with what we are. The handle on a door may be predicted by what I am—its height on the door and its ease of operation, for fingers such as mine. The blankets on my bed—their size, thickness, and the materials of which they are made—may be predicted by my bodily needs.

So may many things be predicted, more or less: the hours of my working day, the ingredients of a meal, the size of a soccer ball, the shape of a boat, the location of a university, among other things. Much of the way that we relate things to one another lies in their *function*.

This applies, too, to many things *non*human. The domestic cat, for instance. The size and shape of its food bowl may be predicted in advance, as may the cat-flap, the flea-collar, the scratching post, and so much more—which are all tailored to what a cat *is*.

Chapter 12

An Interrelated Ethics

We now survey how a philosophy of relations reconciles major approaches to ethics—with the additional step of uniting heart and mind.

THERE IS MORE THAN one approach to ethics. Until now, we have not made any effort to enumerate them and unite them under a single principle. At the same time, we need to unite more than ethical theories. There is the question, too, as to how the *whole* person—opinions, feelings, judgments, sensibilities—relates to cold ethical theory.

Through the discussion of preceding chapters, we already have the means in hand to unite the major ethical theories. We shall consider the "big three": consequentialism (which judges ethics by its consequences), deontology (which views ethics as being self-evident), and virtue ethics (which sees ethics as being rooted in character).

Each of these theories is characterized by the way that we deal with relations between things—that is, the way that we arrange the world in our minds:

- When we judge ethics by its *consequences*, we trace relations between the present and the future.[1] Therefore, an ethics of relations includes consequentialism—though not in every way.

---

1. We should say, the *anticipated* future, since the future may not be as we think it will be.

- When we consider ethics to be *self-evident*, we accept that certain relations exist in this world that are obvious or in plain sight. Therefore, an ethics of relations includes deontology—although it may not always agree as to what is self-evident.
- And when we say that ethics is rooted in *character*, we accept the view that ethics is about one's outlook on the world—which is to say, the way that we trace relations in this world. An ethics of relations, since it comes from within, represents a kind of virtue ethics.

On this last view, it is important to note that virtue ethics differs greatly from *obedience*. When we act according to our character—which is the result of the relations we trace in this world—this has little to do with obedience.

In fact, we all have a natural aversion to an authoritarian ethics, which seems to be artificially constructed, restrictive, and oppressive. Obedience, while it may sometimes be a necessary evil, is a short-sighted ethics, without a vision for the world. We call it legalism.[2]

In the broader context of political systems, legalism tends to lead to authoritarianism—and, more seriously, weakens the role that relations play in understanding and organizing society. It may be tempting for the (post)modern state in particular to impose its authority on its people—there are, after all, so many ways to do it now—yet a de-emphasis of relations presents an imminent danger to all.[3]

Human Compassion

Often, there is little room for *personality* in our ethics. Instead, our ethics seems theoretical, dispassionate, and once removed. Much of our ethics today is about ideals, norms, rights, legislation, and the like. One would be forgiven for thinking that we have forgotten human personality, vulnerability, and sympathy. In a word, humanism.[4]

It is a criticism of Immanuel Kant that Kant's ethics, being based on "pure practical reason," is without heart. Libertarian Jason Kuznicki

---

2. In philosophy, legalism is often used in the narrower sense of the Chinese school of Fajia (E. Harris, "Legalism"). Here, legalism is intended more broadly.

3. For this reason, an authoritarian state cannot endure. It comes up against the limits of sustainability, by de-emphasizing relations that are balanced and broad.

4. I owe this emphasis to my late wife Dr. Mirjam Scarborough, who emphasized the immediate, personal needs of those suffering religious persecution.

summarized Kant: "Our duty is ultimately an impersonal one, a duty owed to reason alone, and neither to self nor others,"[5] while the philosopher and playwright Friedrich Schiller argued that Kant's ethics overemphasized dignity at the expense of grace.[6] Morality, he wrote, should be "second nature."

All around us, we find both *rational* ethics and (let us call it) *natural* ethics. On the one hand, we find a deliberate, principled, calculated ethics, while, on the other, we find moral actions that we do spontaneously, whole-heartedly, and voluntarily. Human action springs not only from precepts but from the heart.

How, then, do we reconcile the two? This was, famously, the problem of the Jade Emperor, who was caught in the tension between his responsibilities and his inclinations.[7]

While one might argue that a natural ethics does not qualify as any ethics at all (it is not suitably articulated or formulated), ethics represents so much more than mere precepts.[8] A person represents a whole world within—mental, emotional, and (some would say) spiritual. This is manifested in many human ways: through joys and sorrows, weaknesses and strengths, follies and virtues, hopes and fears.

In the words of the philosopher Martin Buber, another self is not merely an "it"—"he is not a thing among things, and does not consist of things"[9]—he is not a number, say, or an entity, an abstraction, a cog in the machine—but represents "Thou."[10] There is *personality* there—which one dictionary defines as the characteristics or qualities that form an individual's distinctive character.[11]

---

5. Kuznicki, "Kantian Case for Libertarianism."

6. Schiller wrote, "In Kantian moral philosophy the idea of duty is expounded with such severity that all the graces are scared away" (Curran, *Schiller's "On Grace and Dignity,"* 120).

7. Van Leeuwen, *Narratives of Kingship*, 88–97.

8. One might argue, conversely, that a *rational* ethics is no ethics at all, where one needs to *structure* one's passions.

9. Buber, *I and Thou*, 8.

10. Buber, *I and Thou*, 3.

11. Peter Strawson considered that "personhood is an irreducible subcategory of certain particulars" (Mautner, "Strawson," in *PDP*, 544).

> Our feeling is undoubtedly earlier than our intelligence,
> and we had feelings before we had ideas.
>
> —JEAN JACQUES ROUSSEAU

The Whole Person

We may reconcile a rational ethics and a natural ethics as follows. We have said that, as we trace relations, we need to think not only on the world we see around us but on the worlds inside other people's minds.

We have said, too, that the presence of others in our world adds a quantum leap of complexity to the task of arranging our world, since we now combine worlds with worlds—tens, thousands, even (in certain ways) millions of worlds in other people's minds. Then, too, this is all mediated through semiotic codes.

Now we may note that the involvement of others in our world, and the ways in which they communicate with us, means that ethics may often come down to something all too human. In order to understand these others, we need to be sensitive to their gestures, postures, expressions, and a great variety of semiotic codes—and again, the sufferings, desires, and hopes that lie behind them.

We need to understand—to borrow a phrase from the polymath Thomas Browne—"the motto of our souls."[12] This is a delicate art, which we cannot master without personal insight and feeling.

The economists Samuel Bowles and Herbert Gintis referred to the human species as *Homo reciprocans*: the cooperative actor[13]—the one who acts in synthesis and sympathy with others. One might suggest *Homo reportans*: the one who re-ports, or "carries back." I carry back the worlds of others into my own. I carry their hardships back into my comfort, their innovations back into my tradition, their distress back into my complacency—to offer but a few examples.

If I lose such rapport or re-porting, then moral wholeness is inevitably compromised—because moral wholeness, above all, depends on thinking that is expansive and integrative, balanced and broad.

---

12. Browne, *Works*, 1:102.
13. Bowles and Gintis, "Homo reciprocans."

A grand integration of ethics, therefore—of major ethical theories, on the one hand, and duty and inclination, on the other—comes down to the simple, low-level concept of relations between things and how people *communicate* such relations.

Excursus

One might ask, at the end of this tour, "What now is the *purpose* of ethics, in terms of this philosophy?" Is it happiness, virtue, peace of mind—or any of the many goals that have been proposed in the past?

Purpose assumes that we have a reason for doing something, while the best we can say here is that ethics deals faithfully with relations between things. This does not fully constitute a purpose, yet it *leads* us to the goal.

In chapter 8, we said that we must begin "just as I am." We cannot, so to speak, fast-forward to grasp hold of a completed ethics. Ethics "grows on us." The Islamic jurist and poet Jalal ad-Din Muhammad Rumi wrote, "As you start to walk on the way, the way appears."[14]

---

14. Raturi, "What You Seek," para. 2.

Chapter 13

Esthetic Relations

> Both ethics and esthetics are about tracing relations between things. A philosophy of relations describes the nature of esthetics and its relationship with ethics.

PLATO DESCRIBED ESTHETICS AND ethics in one breath as "the beautiful and the good,"[1] while Wittgenstein wrote, "Ethics and aesthetics are one."[2] A mere cursory look at the relationship between the two suggests that they must have something to do with each other. Art is endlessly portrayed as being morally exemplary or morally corrupt, as if it were an ethical pursuit[3]—while human behavior is often characterized as being beautiful or ugly, as if it were art.[4]

"What is it that distinguishes judgments as esthetic?" asked the philosopher Nick Zangwill. "What do they have in common? And how do they differ from other kinds of judgment?"[5]

---

1. Plato, "Lysis," in *Dialogues of Plato*, 1:66.

2. Wittgenstein, *Tractatus*, §6.421.

3. Called "esthetic moralism," one applies moral categories to art (Peek, "Ethical Criticism of Art," §1).

4. The philosopher Frances Hutcheson, among others, saw esthetics as an analogy for ethics (Walschots, "Moral Sense Theory," 31). The New Catholic Encyclopedia defines ugliness as "the perversion of the characteristic function of anything or anyone" (Lynch, "Ugliness").

5. Zangwill, *Aesthetic Judgment*, §4.2.

Whatever Wittgenstein might have meant, both ethics and esthetics have everything to do with relations between things. Both have to do with the tracing of relations in a broad and balanced way.

Imagine a couch in a room. It is elegant in the balance of its proportions, the texture of its cloth, and the harmony of its colors. Yet as I raise my eyes from the couch, I see a cheap painting on the wall, hung at an awkward angle. I see a chandelier with half its pieces missing and a glaring striplight mounted next to it. Edwardian wallpaper has come off one side of the room and has been replaced with luminous paint. The more I take in the appearance of the room, the less it is esthetically pleasing.

Esthetics may fail, therefore, if the whole is not in balanced relation.[6] In this respect, there can be no fundamental distinction between ethics and esthetics (we shall continue to speak of art or the arts).[7]

> When we analyse the picture into a large number of particles of paint, we lose the esthetic significance of the picture.... The essence of a picture (as distinct from the paint) is arrangement.
>
> —ARTHUR EDDINGTON

Art is about arrangements that are pleasing to the mind, whether this should apply to a couch, a room, fine art, music, dance, or even the arrangement of my garden. The *whole* of the subject under consideration should reveal a thoughtful relatedness.

One may, of course, point to *contested* art. For example, Stravinsky's *The Rite of Spring*,[8] Jackson Pollock's *Convergence*, or Frank Gehry's Lou Ruvo Center. Yet although the works that they produced may not be pleasing to all, the creators trace relations of breadth, depth, and richness, both in their art and in their person.

In fact, we say of anyone who has the ability to appreciate such relatedness that they have culture and cultivation. The masters of any art are

---

6. Balance may include opposites, too—as in the case of juxtaposition, for example.

7. There are many art *forms*—among them painting, dancing, sculpture, music, and poetry, to name but a few. Philosophy professor Yuriko Saito noted that we "unduly limit the range of esthetic issues" (Saito, "Everyday Aesthetics," abstract).

8. Sometimes referred to humorously as *The Riot of Spring*.

set apart by the breadth and depth of the relations that they trace: visual, auditory, tactile, technical, historical, cultural, and spiritual, among other things. Christopher Wren, Édith Piaf, Wolfgang Amadeus Mozart,[9] and Grandma Moses serve as just a few examples of people who had this special gift.

> ### Excursus
>
> But is not *everything* about relations between things? How, then, should we distinguish between anything at all—let alone esthetics and ethics?
>
> We may distinguish both ethics and esthetics from other fields as follows: both are about human perspective, as opposed to the observation and measurement of physical properties. This may still seem to be too broad—yet esthetics *is* broad. Sport, cooking, bridge-building, even politics may be described as artistic.
>
> While we may try to separate ethics and art for various reasons, it is not easily done. There is rather a continuum between the two. This makes it a somewhat arbitrary distinction, which seems hard to decide, and perhaps we should not try.

Art and Ethics

Because art is an ethical practice, it is an expression of our personal values. Just as we reveal our ethical orientations through the way that we behave, so we reveal them through the way that we do various things artistically: arranging a banquet, laying out a garden, organizing an office, writing, painting, or acting.

---

9. The philosopher Derek Parfit called music "the lost battlefield and graveyard of most general aesthetic theories" (Parfit, *On What Matters*, 1:85). In this chapter, music is incorporated in a general scheme on the basis of the relations it traces: musically, culturally, historically, intellectually, and so on.

If one must draw a line between ethics and art, it is this: the artist does not, as a matter of course, *participate* in that which he or she depicts, so that the artist's intentions may often be separated from the art.[10]

To put it another way, art is dependent on a wider context for its meaning.[11] If, for instance, one discharges a gun on the main street, one does not need to ask after the context. It is known immediately. It is a matter of public safety, and the shooter will likely face arrest. Yet if one portrays the same act in a work of art, it could have multiple meanings.

A play about optimism (for instance, *Man of La Mancha*), a painting about anarchy (Briullov's *Sack of Rome*), or the raising of a victory flag on Mount Suribachi may have different meanings in different contexts.[12]

This has a direct bearing on the *regulation* of art as morally acceptable or not. Art cannot be seen at face value. To use the example of the shooter in the street, one cannot confuse the artist's portrayal of the shooter with the shooter themselves—or anything else, for that matter.

Yet there is a bigger reason as to why we should be careful about regulating art. The danger of regulating anything at all is that this may remove from us the ability to relate all things in a balanced and broad way. The regulation of art may skew the relations that we trace in this world.

If one should discourage art that depicts poverty, or oppression, or decadence, among other things, one could minimize or marginalize a tracing of relations that is vital to the dynamics of our society. Art needs to be *seen* which reveals relationships that are vital to people's understanding of the world.

Why, then, has art at times been condemned, ridiculed, or destroyed—and why has it so often been the target of totalitarian regimes?

The explanation surely lies in an observation of Plato, more than two thousand years ago. Plato took the view that art may cause people

---

10. When we understand esthetics in terms of the arrangement of relations, we may reconcile intentionalism and formalism. What is it that settles the *interpretation* of art? Is it the intentions of the artist or my own subjective impressions? In fact, it has to be both. On the one hand, the various forms of art represent a tracing of conceptual relations *for me*—on the other hand, if the artist expresses herself about her art, this is bound to influence the way I trace relations *through* her art.

11. This is the opposite of what is often claimed, namely, that the context of art is narrower.

12. In fact, by characterizing these works of art like this, one may immediately be wrong-footed.

to imagine alternative realities, and that this could have distracting and destabilizing effects on society.[13]

In terms of a philosophy of relations, this is easily explained. We have said that the relations that we trace in this world have everything to do with *motivation*. When we hold up our arrangement of the world to the world itself and there discover a disjoint, we are moved to act. Similarly, the arts may hold up an arrangement of the world to the world itself—so setting the stage for action. In this, one understands how esthetics may result in political, spiritual, or philosophical impulses.

It is often for this reason that people extol certain works of art and disapprove of others—or dissuade certain depictions of, among other things, women, power, violence, or race. We know that they shape our outlook and actions.

Abstraction

For centuries, the notion of what art *is* seemed to be fairly much settled—namely, art was *representational*. One painted recognizable themes, such as trees in a field or women washing clothes.

This changed with the arrival of the *abstract* arts, arts that lack the conventional marks of imitation, style, virtuosity, and meaning. This makes the abstract arts an awkward fit with the figurative arts that we always knew—if any fit at all. That is, the abstract arts lack the so-called *universal signatures* of art.[14]

Some see abstract art in Islamic art, with its nonfigurative, nonrepresentational patterns. Some see it in the splashing ink method of twelfth-century Chinese art. Some see it as going back to *prehistoric* art—perhaps as far back as humans were capable of abstraction.

Having said this, from about the end of the nineteenth century, there was a marked departure from reality, and this became a groundswell in many fields. Wassily Kandinsky signaled the turn in fine art, when he created some of the first abstract paintings around the beginning of the twentieth century—paintings that no longer sought to reflect physical or concrete existence but shapes, colors, forms, and

13. Pablo Picasso said, "No, painting is not made to decorate apartments. It's an offensive and defensive weapon against the enemy" ("Pablo Picasso 1881–1973").

14. Insofar as one may speak of "universal signatures." These are not cast in stone. "Abstract art," wrote artist Peter Roberts, perhaps too radically, "makes no identifiable reference to the external world" (Roberts, *George Costakis*, 41).

gestural marks.[15] Similarly, Arnold Schoenberg signaled the change in music, John Osborne in theater, Walter Gropius in architecture, and others in various fields.

The surge of abstract art—as popularly understood—is especially puzzling, because, before its time, it did not exist. The timing of its appearance, too, is striking, around the beginning of the twentieth century.

Various thinkers have speculated about its cause: Theodor Adorno, Fredric Jameson, and the post-Jungians, among others. These thinkers did not, however, give any thought to the most sweeping change in society at this time. For the first time in history, society became for the most part urbanized and so came to be characterized by a more manageable, built environment, rather than a tangle of almost infinite relations, as one finds them in the natural environment.[16]

Humans were now surrounded by human works—among them, chiseled edges, geometric shapes, and standard sizes, not to speak of artificial timekeeping, pervasive climate control, and unnatural lighting, all of which represented a loss of contingency in their lives.

Abstraction, through its departure from reality, may promise to set us free from human works.[17]

In the midst of a constructed world—a *controlled* world—it delivers an ocean of impressions: the freedom of an abstract painting, the randomness of sculptural machines, or the whimsy of abstract theater. It does not seem surprising, then, that more or less random visual spaces have gained in popularity as human control has increased.[18]

According to the abstract gallerist Christelle Thomas, abstract art "pulls us out of our usual mental state, offering us a chance of at least momentary transcendence."[19]

15. One may also mention Adolf Hölzel, a notable early abstract painter who slipped into obscurity.

16. Among other things, a loss of meaning has been advanced as a reason for the rise of abstract art—however, it seems difficult to correlate abstract art with events that might have brought about a loss of meaning, such as World War I.

17. Artist and psychologist Jeremy L. Dyer, a part-reviewer of this book, suggested another way of putting this. In a built environment, the interpretation is not mine, but my ability to interpret abstract art in diverse ways returns meaning to me.

18. In the Torah, God commanded Israel not to use a tool on an altar, or they would profane it. This symbolized the rejection of human works and may elevate abstract art, in a sense, to the level of the sacred.

19. C. Thomas, "Abstract Art Images," para. 7. Fine art, wrote William Hogarth, leads the eye on "a wanton kind of chase" (Hogarth, *Analysis of Beauty*, ch. 5). This

Esthetic Relations

We may now cast this in terms of relations between things. We have said that our world represents an inseparable net of boundless relations, in which we isolate *regions* of relations in the midst of an endless expanse. This we do through figurative art, which focuses our attentions on particular things—concrete things—and the relations between them.

Now we may note that, even as we reduce the world around us to *regions* of relations, which are often (relatively) simple and simplistic, we *presuppose* the endless expanse in which we found them.[20]

The possibility of imagining this larger expanse suggests that we are able to experience moments where systems of thought or regions of relations dissolve, and reality presents itself to us in its entirety. This we may experience through abstract art, which represents the dissolution of familiar categories.[21]

Thus we may bring abstract art into relation with figurative art: both kinds of art, in a complementary way, have to do with the dynamics through which we trace relations between things. Figurative art traces relations in a bounded context, while abstract art explodes the boundaries, opening our minds to unthought-of possibilities.

There is a similar *psychological* dynamic. We are all familiar with the inspiration that comes to one in the shower, when the mind is in low focus.[22] Amidst the white noise, the random splashes, and the disengagement of one's mind—perhaps even the singing—the imagination may become fruitful. Interestingly, leading scientists have reported that their greatest insights came to them while riding a bicycle, digging a tunnel, watching the sunset, or stepping onto a bus. These are moments where familiar relations between things dissolve, and new ones are formed, and the whole of reality makes itself available to us.

---

would apply all the more to the rise of abstract art, which may set the mind free from the strictures of our calculating minds.

20. Paolo di Sia, a professor of education, contrasts the finite and infinite in various forms of art (Sia, "Describing the Concept").

21. Ferdinand de Saussure used the analogy of a chess game to refer to language—on which Max Black commented, "Then the 'pieces' themselves (one may say categories) are found to be changing" (Black, *Labyrinth of Language*, 28).

22. The computer scientist David Gelernter has done much research on "high focus" and "low focus" thought, which lie at the ends of a spectrum of thought (Gelernter, *Muse in the Machine*, 65–112).

Reconciliation

Finally, in speaking of art, we should not overlook what some have seen as a comparative *absence* of art in certain cultures—including the idea of art itself. Painting, sculpture, or literature, to give but a few examples, may be rare or nonexistent in some cultures. How should we accommodate this in our thinking? Here, our earlier observations about language might help.

We said that there is a *subatomic* world inside words, but our modern theory of language has largely failed to penetrate words' simple atomic exterior. Rather, it has sought to arrest their meanings with carefully crafted definitions—much reducing them, we might add.

The arts may do much the same. They may arrest *concepts* in time. Man-made esthetics may represent a preservative instinct—sometimes helpful, sometimes pathological—to preserve patterns of the past or disseminate patterns of the present, rather than advancing the receptive and creative thinking that the times require.[23]

---

23. The Frankfurt School emphasized an open-ended and continuously self-critical approach to theory (Biddiss, *Age of the Masses*, 219). Karl Jaspers wrote, "Philosophy is the school of [inner] independence, it is not the possession of independence" (Jaspers, *Way to Wisdom*, 119).

Chapter 14

More about Words

In the midst of boundless relations, there is no reason why certain words or concepts should rise above others. Yet they do.

WE NEED TO RETURN briefly to the subject of *words*—which, again, are united with things through associative bonds. Our language, we said, presents us not with a "domino board" but with an endless expanse—and a major *characteristic* of this expanse is that there is nothing given among things that provides us with a stable center. There is no thing, no concept, that takes priority or pride of place in our world.

In fact, our very language is "flat." Our words do not carry distinguishing marks, such as primarius or secundus, to rank them—nor do *things* come with marks that indicate their greater importance or higher priority over other things. They are all of them composed of mere matter or can be reduced to matter, at the end of the day.

Yet it is obvious, too, that we everywhere find what computer scientist Karen Spärck Jones called "term weighting"[1]—also called "weighted words"—which is their document frequency.[2]

---

1. Jones, "Index Term Weighting," 619.

2. Technically, it is referred to as inverse document frequency, for which there are various formulae. The concept here is a little different, in that I write not only about the numerical weighting of terms but about weighting through synotation.

In terms of this metaphysic, there are words among words, which is things among things, that share more relations throughout all relations than is usually the case.

There are at least three areas in which we *weight* words—that is, three ways in which particular words have a disproportiate influence in our language. This happens through

- weighted terms,
- grammars, and
- symbolic language.

These represent major aspects of language and thought, so that it would be a mistake to overlook them in our overall scheme. If we have no means by which to explain such features of language, we may be suspected of having far less than a metaphysics.

Weighted Terms

The word "death" is obviously of greater import than the word "tiddly-wink," or the insult "Idiot!" likely to affect us more deeply than "Clod!" Some words, among all the words which we speak, carry greater weight than others.[3] It is important, therefore, to understand how words are *weighted*—in spite of the philosophical impossibility of weighting words in principle.

Again, a low-level concept helps us to explain it. We said that words are not the *atoms* of language—in the sense that they represent language's most basic, irreducible components. Rather, words have a *subatomic* existence. They contain worlds of relations inside.

Further, if the *synotation* of words overlaps with that of other words, words may *combine* with other words across all the vast expanse of our language and reality. However, some words' synotations may overlap with those of others disproportionately.

For example, the word "death" relates not only to the cessation of life but is related in some way to my ambitions, dependents, investments, legacy—even the seasonal flu, a surge in crime, my erratic pulse, or the broken lock on my front door. Therefore, all of these things potentially conjure up "death," and the influence of death potentially invades each one.

---

3. The same is true, of course, of *various* semiotic codes. We omit this, however, in the interests of simplicity.

A tiddly-wink, by way of contrast, relates to little in our lives—although it is conceivable that it may artificially take on some vast importance.

We may therefore speak of there being special *nodes* in the endless expanse, which are more pronounced than other nodes. At the same time, we may note that these nodes are the creation of our own minds and will differ from person to person and culture to culture.

In some cultures, death is seen as the supreme enemy, while, in others, it is seen as an ever-present companion. In some cultures, personal possessions have a high priority, while, in others, they are commonly shared. In some cultures, time is of great significance, while, in others, it is little regarded.[4] All such variations serve as evidence of our *weighting—subjective* weighting—of things.

Excursus

Why then do we include "weighted words" under a section on personal ethics? What do they have to do with the way we behave?

Their inclusion has to do with ethics insofar as any weighting of words could raise the *value* of such words and concepts. This discussion may explain why we imbue certain words—and things—with value and raise them above other words, where, in truth, there is no rational justification for doing so. Weighted words help us to understand our moral preferences.

In fact, the weighting of words happens in more ways than the overlap of synotations and the creation of nodes. We now further consider how words are weighted through the special emphases of *grammar* and the use of *symbolic* words.

Grammar

A weighting of words of a different kind takes place through our use of *grammar*. Oft-repeated, therefore more weighty concepts are frequently incorporated in our grammars. We find a most basic example in the designation of the noun and the verb.

---

4. One may add issues here, such as gender, space, illness, and more.

The polymath John Herschel included among the laws of nature correlations of properties, on the one hand, and sequences of events, on the other.[5] More recently, our world has been described as a space-time continuum: space in three dimensions, and time in a fourth.[6] We may expect, therefore, that our language will trace relations through the dimensions of space and time.

This is indeed the case, with nouns tracing relations through space, and verbs tracing relations through time—or, in Herschel's terms, nouns referring to correlations of properties, and verbs referring to events. For example, a building is a structure in space, while to build it is a process in time.

This basic and oft-repeated distinction is then *grammaticized*. Instead of creating the distinction anew every time we speak—referring, say, to a garden-*thing* (a garden) and a garden-*act* (to garden), or a walk-*thing* (a walk) and a walk-*act* (to walk), a paste-*thing* (paste) and a paste-*act* (to paste), and so on—we incorporate the distinction in our grammar.[7]

While it would be an oversimplification to suggest that we may apply such a scheme in every case, we may broadly understand our parts of speech in such terms—including adjectives, adverbs, and other parts of speech.

Now within the two categories of noun and verb, some *relations* are oft repeated. For instance, nouns frequently have to do with *possession* (the genitive case), while verbs frequently refer to the *past* (the past tense). Such conceptual emphases, amplified through heavy use, are compressed, for convenience, into *tables* we call declensions and conjugations.[8]

Different cultures repeat different relations more frequently in their speech. Therefore, grammars will differ from culture to culture, because different grammars reflect different cultural traits. In English, for example, one finds the future tense, where in Japanese, one does not; while in Japanese, one finds the honorific case, where in English, one does not. Such emphases typically correlate with features one finds in the culture.[9]

---

5. Losee, *Historical Introduction*, 105.

6. Girifalco, *Universal Force*, 178.

7. This is to say the reverse of what linguists Edward Sapir and Benjamin Lee Whorf proposed. Sapir and Whorf held that linguistic categories influence perception. I propose that perception influences linguistic categories—in this case, grammar.

8. One may add inflection and derivation.

9. *Functional* grammar emphasizes grammatical analysis, which is based on rules that govern social interaction (Halliday and Matthiessen, *Functional Grammar*, 1.3.3).

Further, because we are dealing with compressive techniques, one finds *inconsistencies* of compression in language—for instance, irregular nouns and verbs, and unproductive patterns. In fact, such inconsistencies may aid compression—as we know well from computer programming.

This further has a bearing on the long-standing question as to whether a common structure lies beneath all grammars—so puzzling in their diversity. Max Black noted, "There is extreme variability between grammars."[10] In fact, "Grammar has no essence."[11] This *should* be the case where grammars are not embedded in our DNA but are manifestations of how we trace relations between things in this world.

The same is true of word meanings. Words, rather than being (an Oxford dictionary definition) "an invariant form with an invariant meaning,"[12] are about relations that shift over time, as our *tracing* of relations shifts over time, changing both the words' form and meaning.

Examples of such shift are without number—particularly in cognate languages: "dier" (animal) in Dutch becomes "deer" (ruminant) in English; "Tisch" (table) in German becomes "dish" (tableware) in English; while "petit" and "petite" (small) in French becomes "petty" (trivial) in English.[13] Such change should not be possible with the Oxford definition of words[14] but is perfectly understood where words represent relations within.

Grammar, therefore, is completely a manifestation of things and the relations between them, as we see them in this moment in time, in this place and this culture.

---

10. Black, *Labyrinth of Language*, 62.

11. Black, *Labyrinth of Language*, 64.

12. Strictly, *linguistic signs*, which are words, morphemes, or other units of a language system (Matthews, "Linguistic Sign," in *CODL*, 228).

13. See Steinmetz, *Semantic Antics*, for various examples. Of course, *concepts* change, too. The theologian and philosopher Kevin Hart wrote, "They too have lives of their own" (Hart, *Postmodernism*, 27).

14. Again, linguistic signs. My own language is simplified in the interests of an easier read.

Part II: Personal Ethics

Symbolic Language

Symbols—both object symbols and word symbols[15]—are based on a relationship of *metaphor*: *x* stands for *y*.[16] For instance, a red rose stands for love, a white dove for peace, or a cross for salvation, while a dead cat on one's doorknob is surely ominous.[17]

What is significant to this account of everything, briefly, is that symbols (we shall confine ourselves now to symbolic words) contain uncommonly large networks of relations. The anthropologist Anna Grimshaw noted, "Symbols have the ability to link previously separate areas of conceptual experience and allow the human mind to go beyond what is known or observed."[18] While this begs the question as to what exists that we cannot know or observe, the point is that the more deeply symbolic our language, the more expansive the relations that we trace.

Man's ultimate concern must be expressed symbolically.

—PAUL TILLICH

There has long been speculation as to what symbolic language really *represents*. Is it a door to the transcendent, which we cannot put into ordinary words? Is it a language of belief, which lies beyond statements of fact? Does it serve to disguise something else, such as cultural or political values?

In terms of this metaphysics, symbolic language represents worlds of relations within—relations that encompass far more than our ordinary language encompasses—and so imbues our words with a greater potency, or weight.[19]

---

15. My own language is simplified. There are various kinds of symbols besides, and the same applies to all.

16. As to what symbols *are* is subject to much debate. There is a tendency to speak of symbolic objects in philosophy, and symbolic words in theology. One may also speak of symbolic images, events, names, and more. Paul Tillich wrote that symbols participate in the reality to which they point (Rowe, "Tillich's Theory," 593).

17. This happened to one of the author's friends, who agitated against apartheid.

18. Grimshaw, "Symbol," 836.

19. As to which words are symbolic is a relative matter. Many common words may be regarded as symbolic.

"He is a rock" is something that would be difficult to express without circumlocution; "she's on a crusade" contains a world of history; while "Satan will not win" expresses ideas that could—indeed have—filled books. We may not be able to express such ideas in any other way with the same kind of impact. It comes as no surprise, therefore, that the meaning of symbolic words may take us a long time to absorb or may be difficult to define.

One might then ask: are those relations *real*, which symbolic words trace? They may well be real, if, in fact, symbolic words trace relations truly. Symbolic language, far from being a disease of civilization, as anthropologist Roy Wagner suggested it may be,[20] offers us powers to trace relations that explode the bounds of our ordinary language.

Certainly, *some* symbolic language will be false—yet there is a tendency, therefore, to dismiss *all* symbolic language as false. The fact that *some* of it is false need not mean that *all* of it is false—or anything less than true.

---

20. Wagner, *Symbols*, 1.

Part III

Public Ethics

The third part of this book surveys the field of *public* ethics. This differs from personal ethics, insofar as a *combination* of individuals now needs to reach ethical decisions. The principles of political life are prioritized.

Chapter 15

Political Relations

Public ethics seeks a broad interrelatedness in society. We achieve this through a form of democracy and a broad set of core values.

DEMOCRACY IS OFTEN SAID to be the best available political system. It is, to put it too simply, a system of government by the whole (eligible) population. Even nondemocratic states will give some approval to the idea. Yet there is an idea that is more basic than democracy—in fact, *incorporates* democracy. It is as simple as the *relatedness* of all things.[1]

We need democracy, not merely for the sake of popular sovereignty, political accountability, individual rights, or a host of other things that populate descriptions of democracy. We need it first, because, properly conceived, it is important to the arrangement of our world. Without that, we are all imperiled.

Democracy meets the needs of a healthy *interrelatedness* in society. In a democracy, one elects those persons to democratic office who are broadly representative of the people—so that, when they assemble, they trace relations through all of society, if not the world.[2]

---

1. It is the relatedness of all things, too, that provides the justification for a system of government. It is beyond the ability of individuals to trace relations throughout society or to exercise control over it.

2. Democracy here means, very broadly, a system of government in which all eligible citizens participate in political decision-making. I myself come from a tradition of direct democracy in the church.

However, when we describe democracy in these terms, we see some obvious shortcomings in our *current* conception of democracy.

While democracy rightly guarantees a broad participation in the national debate, it does not deliberately prioritize the tracing of *relations*: for instance, relations between the rich and poor, the built and natural environments, the present and future, or the local and global. Imbalances in areas such as these hold dangers not only for the present but for the future.

Further, in a democratic state, *people* are often prioritized over relations—and so, too, the weaknesses we typically associate with people: populism, personal loyalties, polarization, fleeting fears, vested interests, prejudices, short-sighted thinking, and obduracy,[3] among other things. For good or for bad, democracy is a people-focused enterprise.

When we consider public ethics, therefore, *relations* ought to be the without-which-not (the *sine qua non*). As humanity's influence on the planet grows, we are no longer able to absorb mistakes that may happen through the prioritization of people over relations.[4]

Thus a democracy rests on a low-level concept, far more than it does on the high-level concepts with which we commonly associate it. It rests on relations between things that are balanced and broad.[5] In fact, if this low-level concept is overlooked, an entire political order may possibly malfunction.[6]

This further influences the way that we think about democratic *representatives*—that is, the representatives of the people.

In order to be in a *position* to trace relations through all of society, representatives ought not to be too far removed from their original contexts. If they are isolated from their own society through, for instance,

---

3. Obduracy is a problem of leadership theories that may be grouped under "transformational leadership." Leadership theorists Halcomb et al. summed up *Christian* transformational leadership as "total obedience to the God-inspired vision" (Halcomb et al., *Courageous Leaders*, 217). This has much in common with secular transformational leadership.

4. Statutory law as we know it may often be too weak, and too limited in scope, to bring about the arrangement of society that is needed. It needs to be more expansive and effective than it is today—which refers not to the size of the law code but to its scope and power. "Positive" religion (which emphasizes rules and laws) has known this for millennia.

5. Therefore, the primary goal of government watchdogs is to examine the success with which a government brings everything into broad relation.

6. Not only may a democratic system of government be compromised. If democracies fail to find a holistic balance in society—which is radical democracy—one may question whether there is a democratic dynamic at all.

privileged housing, special schooling, elite policing, or private medical care, this may defeat the purpose of bringing all of society into balanced relation. To put it simply, political representatives should not be out of touch with their constituencies.

Excursus

Which then is it to be? Is supreme power vested in the people, or is it vested, so to speak, in relations between things? Do we favor democracy, or do we not?

The goal of democracy is not democracy as an end in itself but a broad *interrelatedness* of all things.[7] While democracy means "power to the people," such power must be vested not merely in the people but in the arrangement of the world, as it were.

Again, for the sake of a name, we shall call this form of government a *syntocracy*—from the Greek *syn*, "together with," and *krites*, "power"—a form of government in which all things are brought together with all things in balanced relation, *through* the people.[8]

Relational Foundation

It would be helpful to cast our minds back over history, to survey a major *shift* that has taken place

Plato, in the fourth century BC, proposed that the minimum state would consist of just four or five persons: the farmer, the builder, the weaver—perhaps the shoemaker—and one or two more.[9]

Today, with few exceptions, society is vastly more complex, weaving a tangled web of relations through an unimaginable variety of human roles and physical infrastructures.

---

7. Much is made of Plato's philosopher-kings. Yet these were appointed for their knowledge, while being separated from the workers (Menahem, *Historical Encylopedia*, 1:210). This is fundamentally contradictory.

8. A syntocracy, however, should not imply the "iron cage" bureaucracy of Max Weber (Weber, *Protestant Ethic*, ch. 5).

9. Plato, *Republic*, 369.

"The more primitive [that] societies are," wrote the sociologist Emile Durkheim, "the more resemblances there are among individuals."[10] Conversely, as society advances, we are surrounded by ever-increasing knowledge, specialization, and diversity, while our individual competencies cover ever-smaller regions of relations, relative to the whole.

Take, as an example, the *builder* of Plato's state. It is quite likely that he would have known something, too, about farming, weaving, and making shoes. Today, however, the builder is unlikely even to know everything about building, let alone the rest. Even in the simplest of situations, building now requires surveyors, architects, engineers, estate agents, bankers, lawyers, and more. In the case of large construction projects, the number of complementary roles is legion. The same may be said of all major industries.

Not only this, but the *distance* between people has often grown—both physically and socially.[11] No longer does my whole world lie on the doorstep of my own thatched hut. In fact, we often do not know our very own neighbors. As we develop a more complex social diversity, we are no longer able to see into other people's worlds, to understand each other's lives, or even to be aware that they exist.[12]

People play games in lighted rooms, wrote the social thinker John Ruskin, without knowing the losses and gains far away among the dark streets.[13] In short, our loss of interrelatedness in society—and the world—opens up the possibility of a host of ills.

We may not know or may not care whether those worlds that lie beyond our own conform to our standards of justice and equity or function in ways that will safely sustain both their society and ours.

With increasing diversity and increasing distance from each other, there is a far greater need today to coordinate complementary roles and understand complementary interests.[14]

10. Durkheim, *Division of Labour*, 105.

11. Perhaps the most notable example is the gap between *rich and poor*, which reflects other forms of distance. In the USA, as one example, "over the past 40 years, the gap between rich and poor communities has increased dramatically" (Reuell, "Cities' Wealth Gap.")

12. The sociologist Nathan Glazer observed that twentieth-century nations frequently displayed a mixture of ethnic and social class stratification, with the one serving as a surrogate for for the other (Glazer, "Why Ethnicity?").

13. Ruskin, *Genius*, 245.

14. According to Emile Durkheim, the goal of modern societies is to strengthen social integration—which speaks of tracing relations ("Durkheim and Social

> **Excursus**
>
> But are we not so much *better* networked today? Has not the distance between us *shrunk*?
>
> In fact, our networking may *increase* our distance from one another. In the same way that it enables us to connect more broadly, so it enables us—often *steers* us—to join with like-minded groups, to the exclusion of others.
>
> Further, the *methods* of our communication, by their very nature, may not succeed in diminishing (for example) class differences, wealth inequalities, or access to justice. In an important sense, "the medium is the message."[15] Each communication platform advances certain *type*s of communication.

Political Freedom

What, then, of political freedom, if we prioritize the relatedness of all things? Does not freedom, then, take second place to organization? In particular, the freedom of the individual?

Freedom has many meanings. The meaning that concerns us here originated in the eighteenth century with the French *philosophes*.[16] The philosophes considered that society and state exist to preserve the individual and his or her liberty.[17] The individual was therefore "prior to" the group.[18] This idea is still buoyant today.

Yet the freedom of the individual may not guarantee the *good* of the individual—or of society as a whole. Some counterproductive freedoms may be painfully close to home: for example, personal appetites, ways of

---

Integration," para. 1).

15. McLuhan and Gordon, *Understanding Media*, 25.

16. Not all have agreed with the *philosophes*. The philosopher Bernard Bosanquet, as an example, proposed "a reversal of classical liberal priorities" (Blackburn, "Bosanquet, Bernard," in *ODP*, 47).

17. Before the *philosophes*, Baruch Spinoza wrote that "the true aim of government is liberty" (Spinoza, *Chief Works*, 1:259).

18. LaBoy, *Table Matters*, 132.

life, or earning capacity, to give but a few examples. Depending on how a society deals with these issues, much harm or good may result.[19]

Further, we often apply our laws to the behavior of the individual, rather than surveying the larger whole. Behaviors that have devastating effects on the whole may be treated merely as isolated cases, with less than the seriousness they deserve: damage to the national infrastructure, harm to endangered species, or government corruption, among other things.

The theft of a railway line is not to be equated with its value in steel. Harm to endangerered species goes far beyond the loss of isolated organisms. Corruption does not merely mean the enrichment of a few. The wider consequences may be huge and sometimes swift in their effect.

Where one takes account of relations as a whole, lawmakers will have proper regard for the wider impact of individual behaviors and their seriousness in the context of the whole.[20]

> That which is not good for the hive
> cannot be good for the bees.
>
> —MARCUS AURELIUS

Finally, under freedom, we need to consider *property* rights—if for no other reason than the *pervasiveness* of the issue. It is, at one and the same time, one of the most important and one of the most disputed aspects of our common life.

In terms of a philosophy of relations, our priority is the healthy interrelatedness of society as a whole. With this in mind, property rights—both individual and collective—may not in all cases bring about society's highest good (the *summum bonum*). They are not written in stone.[21]

---

19. Bernard Bosanquet emphasized that one should not forfeit community values at the expense of individualism (Bosanquet, *Philosophical Theory*, 12). The philosopher-diplomat Niccolò Machiavelli wrote that states which are in good order will put the common welfare above selfish pursuits (Hanasz, "Common Good in Machiavelli").

20. Of course, there are exceptions, most obviously where individual behaviors do not negatively affect the whole. This includes behaviors that fall under *identity*, to be covered in ch. 19.

21. In fact, it is arguable whether *any* rights can be written in stone, where the healthy relatedness of all things has priority.

Yet one may say this: enduring property rights may be essential in many respects to promote both order and progress in society.[22] This applies not only to land ownership or residential real estate but to property of all kinds.[23]

If my workshop should be taken away from me, my career could be disrupted—not to speak of my life and the lives of various others. If my computer should be snatched from my desk, a project could stall, and a network of connections could be lost. In fact, if even a pen should be snatched from my fingers, some small kind of progress could be delayed.[24] Whatever a society should decide to do with property, "order is the fundamental value."[25]

One should not assume, however, that property rights immediately represent a reasonable benefit to all. While many people own property today in terms of their rights, these rights may be only minimally realized, and property may not be the benefit that it really should be. For example, residential real estate may have high overheads in travel, maintenance, or utility costs.[26] While a right would normally be seen as a benefit, it could be far from it. Ultimately, the issue is human flourishing.

Common Compassion

When we paid our first attention to personal ethics, we noted the presence of *others* in our world, who do not merely modify our game but present us with a world of semiotic codes (see ch. 10). Not only do others feature in our lives as entities that stand in some practical or tactical relationship with us, but we have to take account of them in a *personal* way.

Now we may apply this to public ethics, too. The healthy interrelatedness of society is about insight and feeling, not only towards individuals but towards communities.

---

22. Compare John Locke, "The great and chief end of men uniting into commonwealths and putting themselves under government is the preservation of their property" (Locke, *Two Treatises of Government*, 101).

23. Even where societies advocate inviolable property rights, it is common for property to be shared in various ways.

24. One should not overlook *time* thieves, of whom there are many, among them, inattentive officials, idling employees, or, simply, undependable friends.

25. Miller, "Kirkpatrick Theory."

26. Depending on what housing rights specify, homes may, for example, have no water, security, or privacy.

Part III: Public Ethics

The *collective* life of others—of whole communities, even nations—is vital to the well-being of humanity. To cast this in more ordinary terms, there should be a sensitivity to the "soul," not only of individuals but of entire communities. All sectors of society should have a lively feel for all other sectors of society.[27]

Here, we need to make an important distinction. It is important to think about our social order not only in terms of future *ideals*—because future ideals are more practical, tactical—but to open our hearts to our present *experience*.

As we think on the experience of others—better still, *share* their experience[28]—the world as we know it changes—as does the world that we expect to see. This creates a *visceral* response, as described in chapter 10. We begin to feel what others feel—or, in some cases, as they *ought* to feel. This brings us into sympathy with them and moves us to action—both in the present and for the future.

The polymath Albert Schweitzer might have gone too far when he wrote that all ethics must take root in compassion.[29] However, no ethics should be without it.

We turn, finally, to a nation's *leaders*. Based on the need for fellow feeling, these need to serve not merely as functionaries, executives, legislators—even visionaries. Rather, they ought to be competent two-way communicators. They should be able to establish rapport, responsiveness, and reciprocation with citizens throughout society—and not merely with citizens but with *communities*.

The requirement, in a nutshell, is embodied in US President Franklin D. Roosevelt's fireside chats. It is in the interests of society to have a personable president or prime minister—not to speak of every public servant.[30]

---

27. The philosopher-politician Confucius taught the necessity of *jen*—which is those moral qualities that make us humane ("Jen," in "Main Concepts of Confucianism," §4A). The philosopher Johann Fichte roundly condemned an unconcern for the common good. (Mautner, "Fichte, Johann Gottlieb," in *PDP*, 198).

28. In some cases, people have been exposed programmatically to different sectors of society, to share each other's experience. An example is Zambia under president Kenneth Kaunda.

29. Free, *Animals, Nature*, 53.

30. Rudyard Kipling wrote of the need for a rapport both with crowds and kings (Kipling, "If," in *Rewards and Fairies*, 56). At the same time, we need to check the ever-present danger of the personality cult. This will go without saying for a leader who prioritizes the healthy arrangement of society.

Chapter 16

Informational Relations

Montesquieu proposed that democracy is underpinned by a separation of powers. This, in turn, is underpinned by informational relations.

DEMOCRACY HAS TRADITIONALLY BEEN safeguarded through the *separation of powers*—which is a separation of the executive, legislative, and judicial branches of state. Such separation reduces the concentration of power and provides checks and balances, which serve, in particular, to reduce the risk of autocratic rule.[1] "Power checks power," wrote the political philosopher Charles de Montesquieu, who developed the theory in the early eighteenth century.[2]

There is, however, something more basic than the separation of powers, which *underlies* the separation of powers. This is the free flow of *information*.[3]

The free flow of information is a prerequisite for the task of arranging the world in our minds. Its availability (or not) may be the difference between an open society and a closed one, between propaganda and truth,

---

1. "It may be a reflection on human nature," wrote US statesman James Madison, "that such devices should be necessary to control the abuses of government" (Madison, "To the People," para. 3).

2. Checks and balances were central, too, to John Locke's earlier political philosophy (Jenkins, "Lockean Constitution," abstract).

3. Montesquieu merely infers this but had to favor the free flow of information for the separation of powers to work.

between inscrutable power and civic participation. It may conceivably be the difference between the survival or destruction of the human race.

If we do not know the state of the world—that is, if we do not so organize our information that we have a firm grasp of *informational* relations—we may be setting ourselves up for disaster.[4] This returns us to the simple ethical maxim of chapter 8:

- *Embrace those relations that exist in the world.*

Knowledge itself is power.

—FRANCIS BACON

The Scope of Transparency

When we think of the separation of powers, we may naturally think of the Capitol, the Houses of Parliament, stately courts of law and robed judges, or smartly dressed police on parade. However, the separation of powers typically comes down to far more ordinary things than this.

Suppose that a local regulator has wrongly suppressed a complaint against one of its members—say, a complaint against a lawyer—a common problem, which is called "regulatory capture." A citizen therefore reports the regulator to the ombudsman. The ombudsman assesses the complaint against a list of rules; explains how these rules apply in this case; and, all things considered, states how the situation will be corrected or remedied.

In this example, a legislative body (the ombudsman) has checked a judicial body (a regulator) and this at a fairly basic level, which is the level of the citizen's everyday experience.

Now let us notice that each step in this example rests on *information*: the citizen's original report to the regulator; the ombudsman's rules; the application of these rules; and, finally, the announcement of remedial action. In fact, regulatory capture typically cannot happen unless the

---

4. For this reason, no authoritarian or tyrannical regime will succeed in the long run. They will lack the information they need to sustain their rule. Cyril of Jerusalem, in the fourth century AD, held that even the great antichrist would succeed for only three-and-a-half years (Cyril of Jerusalem, "Catechetical Lecture"). This remains a widely held belief.

regulator hides its rules, makes up new ones, vanishes files, "forgets" a case, or acts irregularly in some other way.

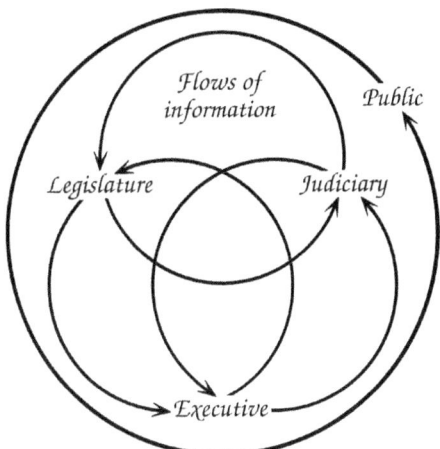

Therefore, the information that we need in order to safeguard checks and balances may be suppressed in quite ordinary ways. These may easily escape our attention: the denial of a receipt, the omission of a signature, a few lost pages, or the neglect of daily procedure. Such things may conceal a world of trouble and are often critical to the functioning of the system.

There is an important assumption which we could overlook. That is that information, where it exists, will be taken up by some system.

We discover, say, that the cricket ball has been tampered with; the facts are therefore taken up by the system of cricket, through the umpire. We track down the perpetrator of a murder; the evidence is taken up by the system of the courts, through prosecutors. We find that the president has issued illegal orders; this is taken up by the system of constitutional law, through parliament.

Ideally, this *uptake* of information drives responsible, accountable, and compassionate behavior.

But sometimes, information may be available—even widely known—where there is no system that will suitably take it up. Information must therefore go hand in hand with its uptake.[5] This is vital, above all, to a functioning state.

We return now to our most general need for information, at the highest levels of government.

---

5. Note, therefore, that the *rules* of the system may obstruct information.

This is covered today by government presses or web portals, which share such information not only with the various branches of government but with the public.

In this regard, we may notice something of crucial importance. Where such presses or portals exist, people not only have the *right* to information;[6] they *have* the information—and for those who are digitally connected in our information age, this information lies just seconds away. There is no need to take people's word for it or to make application for it. One *has* it.

Yet such presses and portals provide only limited information—certainly not enough to catch every significant failure in the balance of power. Citizens should have access to *sufficient* information for the system to work, both in its broadest scope and down to the essential details.

We may think not only of the branches of government. Society is a mix of many powers, and all may selfishly seek to gain the ascendancy: companies and cartels, landowners and shareholders, executives and agents, movements and faiths, to name but a few. We need freedom of information in every sphere.

Wherever information is concealed or distorted, power may go unchecked[7]—which is to say, people may be able to gain personal and political advantage that works against the rational organization of society. Where information is concealed, politicians amass personal fortunes, states are captured, crimes are swept under the carpet, diseases are spread, the poor are exploited, foodstuffs are unsafe, prices are fixed, buildings are raised without permissions—and sometimes they fall—and a thousand, ten thousand ills besides.

It is critically important that there should be ways to shed light on rules, plans, processes, and actions—throughout society. We need to know the why, how, what, and how much in every sphere. In fact, we know, through recent empirical advances, that such transparency moderates the arbitrary, exploitative power of individuals.

---

6. Of course, they may have the right to information even where government presses and web portals do not exist.

7. Importantly, *errors* may go unchecked. The cognitive psychologist Steven Pinker noted that "the community as a whole implements the most reasoned decision," if the free flow of information is sufficient to allow this (Pinker, "How to Avoid Falling," 04:42).

Excursus

Yet what of personal privacy, commercial-in-confidence information, or national security operations? How can one call for transparency *there*?

It comes down to this: how much can we *trust* each other with information? In some cases, if we open up information to all—personal, corporate, or national—there are those who may soon abuse it. For some, information may not serve an expansive and holistic purpose. In fact, quite the opposite.

In such situations, there are two things we may say in principle: we should examine the reasons for a loss of trust, and we should minimize the amount of information that remains secret.

In view of the dangers of withholding relevant information, the failure to provide it should be treated as an offense against society, if not humanity—together with other offenses that, we have said, have far-reaching effects on the whole. So critical is information that whoever improperly withholds it should be removed from positions of responsibility and authority.

This applies to government institutions, public officials, professional oversight, civil servants, company executives, and board members, among many others.

A Balance of Information

The *management* of information, too, should be as transparent as the information itself. Information should, as far as possible, not be *mediated*. As soon as intermediaries stand between the citizen and relevant information—or inscrutable machines—information is one step removed.

Parallel to a threefold separation of powers, therefore, it is crucial that there should be a threefold separation of *information*. Duplication is not enough. Alter one copy of two, or suppress or destroy it, and the value of the other may be lost. In triplicate, information is relatively secure.

While it is possible to *distribute* information today, rather than triplicating it, a simple threefold separation would seem essential, or the very management of information may become obscure.

The question now remains: who should *hold* such information? Three prerequisites suggest themselves:

- information should be securely and systematically stored,
- the repositories of information should be independent of one another, and
- they should offer immediate and universal access to such information.

Information, too, should be independent of the *media*. While the media—sometimes referred to as the Fourth Estate—have traditionally played a vital role in keeping society informed of important developments, they do not offer a systematized repository of information and are unsuitable as full sources of information as here proposed. The media, too, may be driven by their own interests rather than the desire to protect the *public* interest.

We need again to beware of totalizing tendencies. Information does not exist to facilitate perverse control. It benefits humanity. It enables a society to get a fix on the way the world is objectively configured.

In fact, unless there are the few who have special access to information and the power to exploit it, transparency is likely to hold fewer dangers than its concealment.[8] President John F. Kennedy warned, "The dangers of excessive and unwarranted concealment of pertinent facts far outweigh the dangers that are cited to justify it."[9]

It is interesting to note that, in early societies, everyone was able to see into the life of everyone else. Teepees were pitched in an oval, wagons drawn up in a laager, and kraals arranged in a circle—such that these were open societies.[10] Yet through the course of history, this has changed, and it has often been to our detriment.

---

8. The concealment of information, because it represents a reduction or distortion of information, poses a risk to the health and survival of a society, and weakens it.

9. Kennedy, "President and the Press," §1.

10. I lived for several years in such a society, as a boy—as an example, in villages where open huts on stilts were arranged in a circle.

While it is impossible now to return to an open society of that kind, it is possible to recreate one of its central features, which is a well-considered transparency.[11]

---

11. In earlier times, there were secrets, too. However, they were fewer than they are today, for the reason that it was not possible to keep certain secrets in a close-knit society.

Chapter 17

Employment Relations

Public ethics has everything to do with employment. Yet our conception of employment today causes much harm.

EVERYWHERE, WE HOLD THE basic assumption that the ordering of our economic life determines our personal well-being. Here, we reverse that notion. Our personal well-being is to the best economic advantage. It is a simple reversal, with far-reaching consequences.

Public ethics has everything to do with labor (work) and employment (paid work). Typically, at least 80 percent of a labor force is employed,[1] so that theories of employment may have a major influence on society.

Yet while labor and employment are central to economics, we may be forgiven for failing to see it in the *definition* of economics, which is "the study of the production, distribution, and consumption of goods and services."[2] Here, the existence of any living beings is merely implied. The central concepts are theoretical and impersonal.

We begin by setting labor and employment in the context of economic theory in general.

Economic theory, in its infancy, assumed that the goal of economics was the growth of income per head, together with suitable shifts in the

---

1. Methods of calculation vary a great deal. Many organizations publish harmonized values.
2. "Economics."

structure of production—for example, a shift from agriculture to industry. There was growth indeed, yet there were, at the same time, growing social and environmental problems.

Over the years, this became a source of great perplexity and led to much revision. Through *welfare economics* in the twentieth century, the *welfare* of individuals moved to center stage, at least in theory—"welfare" being defined as the satisfaction of human wants through the consumption of goods and services.[3]

Yet we know that we need so much more than goods and services. We need freedom, love, happiness, entertainment, and rest, among other things.[4] More holistic successors of welfare economics have since emerged—among them the *capability approach*.[5] This proposes a raft of economic theories to maximize workers' *capability*—where capability is the capability to obtain those things that one highly values.

In short, the capability approach holds that economics ought not merely to address issues of goods and services but help us become rounded human beings in a balanced society.[6] In terms of a philosophy of relations, this is our interest, too.

Employment

While there have been great shifts in economic theory, from the point of view of *employment*, the shortcomings of our economic life loom large. Apart from the damage that has been done by our global economic order,[7] we tend to underestimate the way in which our economic life has brought unnecessary burdens on us all.

Harmful and hurtful circumstances are bound up with our work as an onerous necessity: traffic jams, rotational shifts, red tape, even

---

3. Not that welfare economics was entirely new. The political theorist Thomas Paine advocated a welfare economics in the eighteenth century (Philp, "Thomas Paine," §4.5). More recently, anthropological functionalism emphasized the satisfaction of biological and social needs (Porth et al., "Functionalism," para. 2).

4. Another way of looking at this is that, in spite of achieving the Pareto optimum, the distribution of income may be undesirable.

5. Robeyns and Byskov, "Capability Approach," §2.6.

6. There is merit in approaching this from another direction. We seek rounded human beings in a balanced society, within which economic needs are met.

7. On the largest scale, disparities between Global North and Global South, First World and Fourth World, global creditors and global debtors.

surrendering our children to strangers—fatigue, oppressive environments, micromanagement, and demands beyond our ability to cope. Nor are we free to be excused, to the point, sometimes, of exhaustion, depression, sickness, rage, and even suicide. We are not in an enviable situation.

Consider the word "employment," which is derived from the Latin *implicare*, to enfold. Much of the original meaning is retained: we are *enfolded* by our employment. We are surrounded, enveloped, engulfed.

Thus Alexander von Humboldt wrote that where labor "does not spring from a person's free choice," it is alien to his or her true nature. In fact, long has there been a debate as to whether the employee is merely marketable goods. Karl Marx judged that it is the selling of oneself and being sold.[8]

> Money supplants or dominates many of the traditional social bonds based on family, tribe, community, and nation.
>
> —JACK WEATHERFORD

Such language might seem extreme. Our experience often seems to suggest otherwise. If we are mere commodities, then we would seem, often enough, to be *cherished* commodities: valued colleagues, thoughtfully motivated and graciously accommodated. However, in view of the heavy burdens we have just surveyed—which are everywhere to be found—in many respects, we are in *bondage* to our work and its circumstances.[9]

A fundamental shift in thinking may help us view things in a new way. Suppose that, rather than working for wages, we should be set free to follow a *vocation*—a word that is derived from the Latin *vocare*, to call. Imagine, that is, that we would be *called* to labor, in the sense of receiving an inner, even divine *summons*.

If we think of a person's labor as their calling, we take account not only of the needs of the marketplace but of the needs of each individual.[10]

---

8. Marx, *Capital*, ch. 6.

9. In a certain sense, we are enslaved—bearing in mind that the meaning of the term *slavery* has shifted and *is* shifting (Zweynert, "Modern Slavery").

10. The educator and philosopher John Dewey spoke of translating people's capacities into their "social equivalents," which means that instincts or tendencies are granted a place in society (Dewey, "My Pedagogic Creed," 77). While it may be hard to

In terms of *relations*, we not only relate the worker to the labor market, but we relate their entire *being* to their calling. This is the ultimate relating of who I am to what I do.[11]

In keeping with this, workers would be provided with everything that serves the end of their calling. They would receive, on the one hand, what we may call negative freedoms (in Latin, *removens prohibens*), which set them free from the need to maintain a home, grow their own food, mend their own clothes, and so on. On the other hand, they would receive what we may call positive freedoms—which set them free to rest, augment their skills, broaden their influence, and more. Further, they would be assisted in dealing with any disability or sickness.[12]

The viability of this approach has long been proved through religious institutions in particular. For centuries, religious movements have set their clergy free from secular pursuits, to follow a calling. Their stipend is then said to liberate them from secular employment.[13]

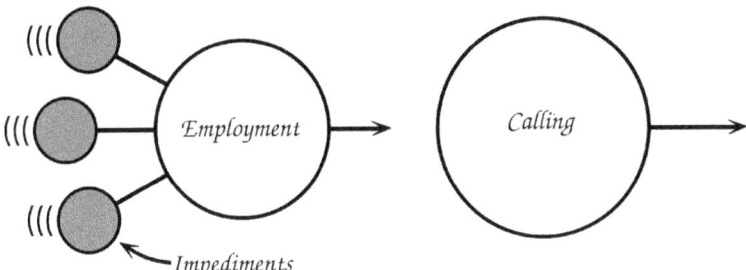

The goal of society, then, is to remove impediments to its citizens' callings and the burdens that they experience in carrying them out. This is already the case—has always been the case—in elite recruitment. It ought to happen in all recruitment. One should offer the employee a *calling* from which impediments and burdens are thoughtfully removed.

---

see some occupations as callings—say, bricklaying or refuse collection—this is easier when we view them in their historical and social context.

11. The social theorist Claude-Henri de Saint-Simon held that all members of society should develop their innate potential—which suggests a calling (Saint-Simon, *Political Thought*, 78).

12. Notice that the *calling* therefore has a greater authority, as it were, than the employer. It may also serve to unite employer and employee around a common purpose.

13. The religious manual *Evangelical and Congregational* notes, "A church does not employ a minister. It releases him from the need of secular employment and undertakes properly to maintain him and his family so that he may do the Lord's work among them" (Booth, *Evangelical and Congregational*, 21).

> ### Excursus
>
> Yet can we ever hope to mesh millions of callings with society's uncountable needs? Moreover, how should we accommodate more *fanciful* callings—the would-be poet laureate, say, or the surfing champion?
>
> Here we need to set aside a common misconception, namely that a calling emerges from *within*.[14] This notion unnecessarily complicates the idea.
>
> Rather, a calling comes from the *outside*—and is fulfilled when it is understood or accepted within. Even if a person should not sense a personal calling or discern it, the *assumption* that one is dealing with a calling makes the vital difference. By viewing employment as a calling, we have a different way of thinking about the individual in society.
>
> Rather than placing the emphasis on the employment contract, we place it on the whole person. Society becomes not so much a market in which one sells one's labor—often needing to hack one's way through in one's own interests—but is streamlined for the benefit of all.

It is not hard to see that the liberation of the *individual* may liberate an entire population. Individual liberation becomes social liberation.

Suppose that I take an hour to drive to work each day. So, too, do most of my friends. We call it the *commute*—so often a burdensome necessity. Now a new model of employment seeks to liberate us from impediments that surround our calling—through affordable housing, mass transit, job distribution, and so on.[15]

This will not merely be to my own benefit but to *society's* benefit as well. As the problem of my own one-hour commute is solved, larger

---

14. In the religious sphere, a calling lies not so much in us as in the God who calls us, so that we do not so much *present* society with our calling as *receive* it. Even so, it may be worth thinking on how a (perceived) calling may be supported by incentives.

15. The scope of equal opportunity covers "recruitment, hiring, training, layoffs, discharge, recall, promotions, responsibility, wages, sick leave, vacation, overtime, insurance, retirement, pensions, and various other benefits" (Childress, "Equal Opportunity," para. 2).

problems are solved with it. Housing becomes more affordable, labor becomes more plentiful, productivity increases, community health improves—and so much more.¹⁶

Therefore, in the way that one acts towards *one*, one reorganizes all of society—unifying the personal and social aspects of labor. Here, employment as a calling offers us a new form of self-adjusting economics—in fact, a self-adjusting society.

Further, the concept of calling offers a justification for the *services* of the state, among them schools, hospitals, pensions, and police. These all exist in the service of personal and social liberation—and justice.

In the twenty-first century, individualism has brought us to a point where we have considered too much our own needs—whether this be the needs of the individual, the group, or the nation. To consider others' needs is to consider our own. To seek their well-being is to seek our own. Each one matters to the well-being of all.

More than this, we may now interpret all contributions to society as positive. Farmers, builders, investors, bakers, police, and many more should—within reason—not be seen as burdens to society or enemies of the people but as valued contributors to its success.

The Ground Plan

A major aspect of employment is *remuneration*, and a major *problem* of remuneration is its relation to our needs. Such remuneration may be too low—and, arguably, too high. In order to prevent remuneration from falling below the floor of people's needs, it is common practice to set a minimum wage.

However, in terms of a philosophy of relations, the question is not merely how much is required to *sustain* a person but how one's remuneration will bring one into more balanced relation with all of society and the world.

The economist Adam Smith held that, if one leaves people to themselves, the rest will follow. The freer and more general the competition, the greater the advantage to the public. The political philosopher Yves Simon described this as "the belief that the good of the social whole . . . is best procured by the spontaneous operation of elementary energies."¹⁷

---

16. Conversely, take away the notion of a calling, and burdens pile up again.
17. Simon, "The Instruments of Government," in *Philosophy of Democratic*

Yet it is precisely through the spontaneous operation of elementary energies that our economic life swings to extremes.[18] Economics needs to build a more balanced relatedness in society—which, in the case of the minimum wage (an absolute standard), means that we shall favor a *proportionate* wage (a relative standard) instead.

When two acrobats lose their grip of each other, they tumble in any direction. Just so, a lack of relatedness in economics may lead to the extremes that make it necessary to impose a minimum wage in the first place. To put it another way, we need the kinds of controls that bring people into more balanced relation with one another.[19]

Government. Laissez faire ("Leave things alone") is a term that is commonly applied to this view.

18. Another way of putting this is that there are many possible market failures; therefore, an economy is not self-adjusting.

19. The more holistic economics has become, the less one has thought of it as self-adjusting. More recent economic approaches presuppose constitutional guarantees, human rights legislation, and development policy, among other things. This is to bring things into relation more broadly than was the case in the past.

Chapter 18

The Imposition of Relations

*Public ethics are imposed on society by coercive institutions.
Coercive institutions in turn require legitimacy.*

EVERY SOCIETY IS PROTECTED from harm by its coercive institutions, or the instruments of state. Such instruments include the army (for external threats), the police (for internal threats), even traffic police, firefighters, game wardens and rangers, social workers, and so on.[1] All wield, or potentially wield, some kind of power.[2]

The *purpose* of the instruments of state comes down, once more, to a simple idea, a low-level concept. Its aim is to safeguard a broad balance in all things—and to restore that balance where it is lost.[3] It rests on the *broad* view, which surveys our world expansively and holistically.

I am back again, in the room with the table. Again, the table stands in the center of the room. This time, I am surrounded by friends. *Together*, we decide to move the table to the corner of the room.

---

1. These are all referred to as police powers.

2. One may also consider the Critical Theory of the Frankfurt School, which emphasized *insidious* control. The literary theorist Lydia A. Fillingham wrote, "Physical force lies unspoken behind many of society's relations" (Fillingham, *Foucault for Beginners*, 140).

3. Naturally, in the very act of imposing a balance on society, we may *upset* the balance. From this follows two things: we need to establish what true balance is, and we need to use coercion chiefly to maintain that balance.

But now, consider the inevitable. One of my friends declares that, actually, she is not sure where any table at all should stand. Another says that he is definitely opposed to moving it; it should stay in the center of the room. The rest of my friends would like to move it. What, then, will prevent a meltdown among friends?

Given that these friends are equals, yet unable to reach a timely consensus, a decision of the *group* must prevail. This, of necessity, will include a decision-making *process*, and this process will more or less find favor with all those involved.

We may base a decision on an abstract majority (a certain number in favor and a certain number against, for reasons unknown), or some of my friends might, rightly or wrongly, have a *weighted* vote—say, the studious ones, as they would put a table to a higher purpose as a study desk. To put it another way, individual differences of opinion will be nullified—somewhat abstractly, we might add—in favor of a majority. Parallels to the political process are not hard to see.

In the context of a *nation*, then, democratic decisions become laws that enjoy supreme consent, and these laws apply equally to all citizens. According to Yves Simon, "the systematic character of one's preference" now has priority.[4] That is, the *system* has priority over one's own personal persuasion—and if it needs to, *imposes* certain relations on society, through coercive institutions.[5]

Not all coercion comes about through coercive institutions, of course, and even coercive institutions will offer positive incentives for certain behaviors.[6] In other words, *persuasion* may play a larger part than coercion in many cases—and this may often be barely discernible, if at all. Further, society's behavior is regulated by social norms: moral standards, dress codes, and table manners, among other things.

That is to say, society itself brings a pressure to bear on us, which is not codified by rules.[7]

---

4. Simon, "The Instruments of Government," in *Philosophy of Democratic Government*.

5. However, the effects of laws go far beyond the laws themselves. One law wisely chosen might make a hundred redundant. It is not as if laws are always directly coupled with the situations they address, although, of course, they often are.

6. Sociologist Amitai Etzioni identified physical, material, and symbolic types of compliance, brought about through coercive power, remunerative power, and normative power (Etzioni, *Comparative Analysis*, 22).

7. The jurist and historian Friedrich Carl von Savigny distinguished between popular custom, established law, and legal opinion as sources of justice (Mautner

And then, much of our behavior lies beyond any coercion at all. There are things that we do simply because we love to do them, because it is in our nature. We enjoy a meal together, we garden, we play, we even go to work, not because there is any coercion, but because this is what we enjoy. "Against such things," wrote Paul of Tarsus, "there is no law."[8]

Political Legitimacy

While there is a need for coercion on the one hand, it needs to be accepted as *legitimate* on the other. Without a sufficient level of acceptance, laws cannot command obedience. And if a populace begins to see itself as alienated or spurned politically, economically, culturally, or socially, it may become restive—ultimately noncompliant, subversive, or rebellious.

The philosopher Charles Frankel wrote, "Those who submit have to believe that it is for some good purpose."[9] In fact, this applies not only to the laws themselves but to everything *about* them—above all, the institutions that approve and apply them. We refer to such acceptance as political legitimacy.[10]

> Force is to be opposed to nothing but
> unjust and unlawful force.
>
> —JOHN LOCKE

Now consider our simple observation of chapter 10, that whenever we hold up our conceptual arrangement of the world to the world itself and there find a discrepancy between reality and expectation, we are moved to act. The *purpose* of this action is to reduce the difference between the two.

In order for political legitimacy to exist, citizens need to hold up their own conception of the political order to the political order itself and there discover largely what they expect. This, in turn, will rest on their

---

"Savigny," in *PDP*, 501–2).

8. Gal 5:23.

9. Frankel, *Case for Modern Man*, 49.

10. This refers to a govermnent's de facto ability to have its word pass for law, as opposed to whether it has political legitimacy on moral grounds.

understanding of how government works, how the democratic process unfolds, and how laws are applied. We call such understanding political education.

Rightly conceived, political education helps a nation to hold expectations that are helpful to all.[11] Wrongly conceived—that is, if political education is abused—it may degenerate into damaging ideology or propaganda.

We need to note, too, a few possible *pitfalls* of coercive institutions. Above all, coercion may signal the failure of political institutions—in fact, the failure of *society*—to work as they should. Where coercion becomes stronger and more persistent, it may indicate the suppression of a malfunctioning whole. "Frequent use of coercion," wrote Yves Simon, "evidences weakness."[12]

If, for instance, people are incarcerated in increasing numbers, if penalties for crimes become more severe, if the power of the police increases, or if the laws are disproportionately applied—say, to minority groups—this may point to social ills that need to be remedied.[13] In fact, it will point to a dearth of relations that are balanced and broad, and this poses a hazard for all.

Coercion is needed, most basically, because we are not all willing to embrace all the relations that exist in this world, insofar as they possibly can. This is plain to see. People readily sacrifice whatever falls outside the paltry regions of relations that occupy them, so neglecting—for example—ecosystems, social classes, ethnic minorities, and so on. This having been said, we should come as close to a noncoercive ideal as we can.

Motivations for Coercion

We may now turn to the various forms of coercion themselves, above all, crime and punishment, and our rationale for using such punishment.

---

11. This means, too, that there must be a clear distinction between the finality of democratic decisions and a kind of suspension of these decisions for the purpose of continuing debate. Any confusion between the two may undermine democratic decisions, on the one hand, or suppress debate, on the other.

12. Simon, "The Instruments of Government," in *Philosophy of Democratic Government*, §1. "Frequent" would, of course, need to be defined. Simon himself devoted some thought to this.

13. Conversely, one should beware the retreat or resignation of law. A particular hazard is identity politics, where authorities respond with ad hoc concessions, which are merely temporary.

Unsurprisingly, we all feel uneasy about institutional coercion. Even a soldier or police officer does not typically use force on natural impulse but as the instrument of civil justice. Yet there will be actions that are harmful to the public good, where force must be used. If the hive has priority over the bee, then coercive interventions will be required.

This brings us to a core question of penology, which is the *punishment* of crime. Should we think of crime and punishment as a matter of cost and benefit to society or as absolute and timeless?

In terms of a philosophy of relations, the answer is clear. We have said that we act on the way that we have arranged the world in our minds. This means that crime, too, has to do with *mental models*. Therefore, the criminal's potential to develop a new understanding of the world—that is, to trace new relations in this world—will be of critical importance to the way that we manage crime.[14]

A criminal's outlook on the world must change, or the world itself must change—or, perhaps, the *criminal's* world must change—if a crime is not to be committed again. With this in mind, the theologian Clement of Alexandria differentiated between one who "falls into any *incurable* evil" and one who does not.[15] There is a future for the latter, though not for the former.

Traditionally, various rationales for punishment have been put forward:

- it serves as a deterrent (to warn others of the consequences of a crime)
- it is rehabilitative (to help criminals reenter society)
- it is educational (to teach both perpetrator and victim about common values)
- it is incapacitative (to remove offenders from society)
- it is retributive (to harm those who harm others)

There has been an ongoing debate, as to which of these is to be preferred.

In terms of a philosophy of relations, the answer must be: *all* of the above. All of these help offenders to understand social relations in ways they did not understand them before. Arguably, even retributive justice

---

14. The philosopher and social reformer Jeremy Bentham sought to develop the principles of utilitarianism for the treatment of offenders (Bentham, *Works*, 579).
15. Clement, *Stromata* 27.5.

("eye for eye, tooth for tooth")[16] helps perpetrators understand the wider implications of their crime—that is, to trace relations as they never traced them before.[17]

We may add one further rationale to the above. A criminal's conviction helps the public to shape a good understanding of their world. Without a criminal justice system, our understanding of crime would be much reduced. Not only that, but a criminal's conviction helps us understand the seriousness of a particular crime, in relation to the entire system that judges it—in fact, in relation to all of society.

In this regard, how should we relate the seriousness of the crime to the sentence imposed? This includes such cases where the nature of the crime has the potential to seriously damage, even annihilate society. What kind of punishment fits the crime?

Coercive institutions ultimately have power over life and death—they "do not bear the sword in vain"[18]—particularly in times of war or serious threat to the safety and stability of society. This introduces the question of capital punishment.

On the one hand, the factors we may consider relevant to capital punishment—in fact, *any* kind of punishment—seem bewildering. There is a sea of relative considerations. On the other hand, we find something at the core of the question that is little considered. In many societies in the past, there were not the *means*—for example, police and penitential institutions—to permanently restrain those who were a fundamental threat to the life of all, except death. Today, the safety of society remains central—including the safety of those who themselves are incarcerated.[19]

---

16. Exod 1:24.

17. An eye for an eye, and a tooth for a tooth, may therefore be interpreted as representing a broader principle: offenders should come to trace new relations in the world.

18. Rom 13:4.

19. This may become a critical question if we should colonize other worlds such as Mars. We may return to a situation where it is not possible to restrain people permanently who threaten the existence of a colony.

Chapter 19

Relating Identities

Identity was once fairly fixed. Often, one's survival depended on it. Today, shifting and multiplying identities present a special challenge.

HAVING SURVEYED THE PURPOSE of our political order and the reasons for protecting it, we are now in a position to consider the subject of *identity*.

Social psychologist Peter Weinreich wrote that one's identity is "the totality of one's self-construal."[1] In terms of this metaphysics, it is about the way that individuals relate things to things—which is the way that we arrange the world in our minds. It belongs to everything we are and everything we do.

Therefore, identity is *ordinary*.[2] It lies not in great accomplishments, special status, distinctive looks, or any finite set of personal characteristics, but in all the detail of my daily existence.[3] There are four things that follow from this.

- When we consider the issue of a person's identity, we are dealing with an integrated totality—a whole—not merely an assemblage of plug-in components.

---

1. Weinreich, "Identity Structure Analysis," 26.

2. Raymond Williams referred in the same sense to *culture* as ordinary (Williams, *Resources of Hope*, 91).

3. Identity may include *physical* characteristics, therefore. Altered physical characteristics change the way that we relate things to things. Various kinds of surgery, too, bring about a shift in our identity.

- No person traces relations quite the same as the next. Therefore, none of us shares quite the same identity.

- Our identity may *change*—since those *relations* may change, which we trace between things.[4] In fact, our current identity will differ not only from the identity of others but from our very own identity at different times and places.[5]

- The relations that we trace will (ordinarily) change very slowly.[6] This has important consequences, since, if identity is immutable or *practically* immutable,[7] presumably it deserves to be protected by absolute rights.

We have therefore described identity basically and broadly through the low-level concepts of things and relations. This sets the present view apart from other possible views—again, that identity is fixed, that it has a finite set of characteristics, and so on.

At the same time, we find here a *cautionary* thought. We have said that we may trace relations broadly, or we may trace them narrowly. Therefore, our identity may be broadly based or narrowly based. It may include many characteristics, such as socializing, reading, building, mothering, gardening, or worshiping—characteristics that fill out our being in various ways—or it may be narrowly based, so as to include only a few such aspects, to the point of psychopathy.

History of Identity

In former times, the subject of identity was not much considered. A hundred years ago, the very concept of identity was virtually unknown. The reason for this is simple. In former times, for many, there was no other

---

4. Many of the relations we trace are deeply ingrained in us: our physical circumstances, childhood environment, lifetime experience—among other things.

5. Occasionally identity may *switch*: for instance, where one experiences major trauma or receives major insight. Social psychologist Peter J. Burke calls it the "disturbance" of identity (Burke, "Identity Change," abstract).

6. In keeping with this, there will seldom be an abrupt change of *ethics*, since ethics results from the relations we trace between things. Similarly, we cannot swiftly change our conscience or any sense of guilt or shame.

7. Shahram Heshmat, a specialist in the science of choice, describes *schemas*, "the glasses through which we look at the world," as "stable and rigid" (Heshmat, "Overcoming Negative Self-Thinking," para. 2).

race, no other religion, no other language, no other role to play. Further, there was often little choice in the matter. The very survival of the family and of the larger clan and society often depended on fixed identities.[8]

Today, a global mix of cultures has driven the diversification, even proliferation, of religious, moral, political, and other identities—while, at the same time, economic and social necessity has retreated. Thus the importance of identity has grown.[9] This is not to say that identity was not important in the past. However, the issues surrounding identity were, as a rule, not as pressing or as varied as they are today.[10]

The number of identities we now come into contact with is legion: specialist, generalist, straight, gay, atheist, believer, punk, professional, macho, feminist, first nation, immigrant—and so on. Some of these identities may continue from generation to generation, while others pass in and out of existence as the years roll by.

An obvious question now arises. What shall we do with identities? More than this, what shall we do with *conflicts* of identity? Should we aspire to a monoculture? A polyculture? A segregated culture? Do we *need* to be involved in interminable, emotional skirmishes—seeking to defend or advance our own identities, not to speak of those of others? Many have grown weary in the fight.

It is not immediately clear, if we simply survey the facts, whether we should equalize identities, celebrate them, separate them, suppress them, or do anything at all. It is the problem of Hume. An "is" cannot become an "ought." There is no logical reason—no reason under the sun—that we should treat identity in any particular way.

Yet as things stand in our twenty-first century, we have set ourselves up for major conflict.

On the one hand, we have embraced social pluralism.[11] According to philosophy professor Calvin Schrag, this is "diversity rather than

---

8. Nonetheless, there were wars, revolutions, the slave trade, and mass conversions, among other things, to disrupt stable identities. However, the meaning will be clear. There was less imagination, flexibility, or freedom than there is now.

9. Whether one can choose one's identity is a matter of debate. If one cannot, it may be cause for both celebration and dismay. We deal with the subject of free will in ch. 33.

10. The very word "identity" implies the existence of many identities. It is a broad category, which subsumes many variations, even opposites.

11. Some define pluralism as competition between groups and organizations for rewards and benefits in society. If we accept identity as ordinary, the meaning here is broader.

homogeneity, multiplicity rather than unity, difference rather than sameness."[12]

On the other hand, we have adopted the doctrine of absolute rights. In the words of philosophy professor Carl Wellman, an absolute right "always holds, i.e. disadvantages some second party, within its scope."[13]

Such pluralism, wrote the sociologist Ronald Fletcher, makes "the problem of preserving order and freedom very great."[14] That is, on current views, we set ourselves up for interminable wrangling and conflict.

Before we outline a solution, we shall put some basic foundations in place.

Most obviously, the interaction of identities takes place in *society*—therefore, identity has to do with social and political relations, where decisions of the *group* may take priority over decisions of the individual—though not in every case.[15] Therefore, the first question we may ask is "What may be described as public and what private?"

Secondly, a major question, we said, concerns the *immutability* of identity. If identity is immutable, then there is little point in seeking to change it or legislate it.[16] In fact, there is good reason to protect it. In terms of this metaphysics, all identities are either immutable or *as good as* immutable. Race, ethnicity, religion, sex, gender, age—these, and many more, are either unchangeable or unchangeable in the short term.[17]

---

12. Burkhard, *Apostolicity Then and Now*, 150. Kevin Hart considered, "Our world is irreducibly plural." There is no alternative to pluralism (Hart, *Postmodernism*, 85).

13. Wellman, "Rights," 696. In the words of Simon Blackburn, "Rights are frequently held to 'trump' other practical considerations . . . perhaps existing in virtue of more central considerations of the duties we owe to each other" (Blackburn, "Rights," in *ODP*, 331). Absolute rights, or people's interpretation of them, may ultimately lead to civil war (Moyn, "Why Do Americans," para. 13).

14. Fletcher, "Plural Society," 655. It is problematic, wrote the Oxford researcher Alberto Giubilini (Giubilini, "Conscience," §6.1).

15. Above all, certain identities may be considered private and of little or no interest in the public arena.

16. Unless the transformation of society should indirectly transform identity.

17. The relations that we trace in our minds are vast, thus difficult to shift *as a whole* at any speed. Further, there is the tendency—described in ch. 4—to think thoughts "we have to think." There will be exceptions to the rule, but in general, identity is entrenched. Notice that this may explain reeducation camps, which take seriously the difficulty of turning people around. Such camps, however, contradict this philosophy: above all, they reduce the relations we trace.

Again, one needs to beware *totalizing* tendencies. Legislating identity may prove to be perilous, both for society and the environment.[18] Our desire to control identities may reveal totalizing tendencies, which weaken the role that relations play in our wider understanding of the world.

We don't need a melting pot.... We need a salad bowl.

—JANE ELLIOTT

We may now answer the question as to what can be done where pluralism conflicts with absolute rights—which is to say, where there are conflicts of interest between various identity-bearers.[19]

Equality Benefits

We already have widespread measures in place, to assist people where they are disadvantaged through weakness in numbers or weakness of person. Above all, we include all persons in our democratic order and under the rule of law. We promote equality, too, in a multitude of ways: wheelchair ramps, court interpreters, legal aid, flying doctors and paramedics, and so much more. Where someone's identity leads to inequality, we put measures in place to equalize the situation.

Secondly, in many ways, today, we spare people the trauma of *enabling* another identity. This refers to *direct* enabling, as opposed to indirect enabling—the most famous example of which is taxes.[20] In many cases, people have the right *not* to have their identity transgressed in order to sustain another. Most famously, we enact conscience clauses and recognize conscientious objection. In terms of this metaphysics, this

---

18. In fact, the state itself may be the cause of the identities it sometimes seeks to suppress. Certain political systems or social orders will tend to create certain identities without necessarily being aware of it.

19. Contrary to popular belief, we have already found and implemented widespread solutions to this problem. However, this has been rather too ad hoc.

20. My taxes may, for instance, indirectly enable corrupt politicians to travel—but there are more sensible examples. Both the American and French revolutions originated in tax revolts (Burg, *World History*, vi–viii).

is correct, again, on the grounds that identities are immutable. One does not overrule or override immutability.[21]

This should not mean, however, that one may go about thoughtlessly with another person's identity. One should not deliver unwelcome surprises for people—which is *tantamount* to discrimination. We do not wish to have a wedding disrupted, a vacation ruined, or work turned down, because there was some surprise about identity in the mix. Such surprises, great or small, must be removed.

Thirdly, the refusal to enable another person's identity need not *in itself* amount to discrimination, regardless of which side of that refusal I am on. Again, this is for the reason that all identities are immutable.

Once more, a fundamental shift in thinking may help us view things in a new way. Where stand-offs involving identity truly cannot be resolved, and where one party is disadvantaged, it may help us to think not in terms of imposing penalties but of granting *benefits*. We shall call these "equality benefits."[22]

Where there is *loss* for reason of identity—which need not be synonymous with abuse—the state may develop a system of equality benefits. That is, in place of penalties, there may be benefits, provided by the state. Thus losses would be offset by the state.[23] Not only would this avert conflict. It would likely be the less expensive, less disruptive, and less hurtful option than many of the situations we encounter today.[24]

---

21. I have the indelible memory, from school, of a girl who was instructed to read a text but stopped when she came upon blasphemy. The teacher's right in the classroom overrode her (very great) distress.

22. This is a logical outworking of the separation of religion and state, and a rejection of "positive" religion as Hegel called it (Hegel, "Whence Came the Positive Element [in Christianity]?," in *Positivity of Christian Religion*, pt. 1, §4). While this may be contrary to some beliefs, the proposal enlarges freedom of religion, too.

23. By providing benefits to certain people, this should not put them in the spotlight or single them out, so that they should feel they are exceptions to the rule.

24. Naturally, equality benefits would need to be regulated. If every citizen were entitled to equality benefits, for any kind of disadvantage, few would be excluded.

> ### Excursus
>
> Someone might point out that we have *missed* something. Identities are often formed through the course of *history*, so that, *through* the identities we create, valuable lessons are transmitted to later generations. For instance, covenantal relationships, provider roles, or customs of hospitality.
>
> This is important from the point of view that the honor or stigma attached to identity do not arise out of nothing. Sometimes, in the past, they arose from astute observation over centuries. In terms of this metaphysics, they resulted from the long and often careful tracing of relations between things.
>
> There is the tendency today to *sola scientia*, which is to cut off the past and to judge all things on the knowledge and priorities of the present.[25] However, one should be careful of discarding the attitudes and actions of the past. At the least, one should understand why one should discard them. This means first understanding their value in their original context.

After all has been said, certain identities will not serve the common good—or may do harm to all.[26] If identities are ordinary, in the sense that they include everything that defines a person, it stands to reason that some of the things that define a person may do harm to others. There are *malevolent* identities, *destructive* identities, *corrupting* identities, and so on—and in some cases, we may not even be aware of what our identities do. The state will therefore act in the interests of all by curtailing some identities, on well-considered grounds.[27]

Let us now notice what we have done through the course of this chapter. By applying some of the principles of this book, we have defined

---

25. The theoretical physicist Richard Feynman typifies this view: "History is fundamentally irrelevant" (Feynman, "Feynman: Take the World," 8:40).

26. In its quest for social balance, if the state discovers that certain identities compromise this, various kinds of interventions may be required. This should be done not on the basis of sentiment or belief but on the basis of broad and methodical investigation.

27. This may appear to contradict the *immutability* of identity. Identity in such cases is subsumed by the greater need for holism in society.

identity in terms of relations, we have described some of its vital characteristics, we have thought holistically and broadly about the subject, and we have rejected totalizing tendencies—among other things. We have, therefore, in multiple ways, applied the ethical maxims of chapters 8 and 9.

It is a positive ambition, to embrace multiple identities in society. The existence of pluralism affirms the value of all, reveals what matters to our common humanity, enriches the common experience, enables people to serve one another, and enhances the cultural and intellectual richness of society.

Chapter 20

A Boundless Environment

The natural environment, due to its boundless complexity, lies beyond the condition *of things and is forbidden territory.*

INCREASINGLY, THE NATURAL ENVIRONMENT (nature, for short) has become a pressing issue in our modern world. In fact, we have a major environmental crisis—yet all too often, we seem to bring fragmented or piecemeal reactions to its destruction.[1] Above all, we are not agreed on philosophical *grounds* for the preservation of nature.

The father of environmental ethics, Holmes Rolston III, rightly saw that "the epistemic crisis is as troubling as the environmental crisis, and one must be fixed before the other can."[2]

There is a pressing need for a rational basis for a healthy, balanced relationship with our natural surroundings.[3] While *reactive* measures are important, on individual, organizational, and governmental levels, the philosophical underpinnings are sorely missed.

The issue of the environment, more than any other, takes us back to a low-level issue of chapter 7: the issue of *number*. So numerous are the relations between things—in this case, the complexity of the

---

1. Another way of putting it is that we find little operational guidance for defining policies, strategies, and outcome targets.

2. Rolston, "Nature for Real," 41.

3. This includes the four spheres: atmosphere, hydrosphere, lithosphere, and biosphere.

environment—that we have neither the ability to understand them nor the power to control them—not with all the computing power in the world—or all the machines.

The pioneer of philosophy cafés, Marc Sautet, wrote, "Scientific advances create the illusion that men can become masters and owners of nature."[4] But not only that. They create the illusion that men can become *stewards* of nature. This is a fundamental mistake, which lies at the root of many of our dealings with nature.

The natural historian Sir David Attenborough typified this view when he said, "We have the power, we have the knowledge actually, to live in harmony with nature."[5] While he was well-meaning, this is self-defeating. Our power and our knowledge have fundamental limits.

Holmes Rolston III typified the way that we imagine a solution: "We are quite capable of understanding enough [to] survive in a sustainable biosphere."[6] Yet if, as we here propose, we are *not* in fact capable of understanding enough, then it is not through our understanding that we shall find the solutions that we need.[7] That is, it will not be by applying our minds to the problems.

An Uncontrollable Sphere

Since issues of nature are issues of *ethics*, they relate to a number of our earlier maxims, in chapters 8 and 9:

- *Think expansively and holistically.* If our thoughts will range through all the world, this is sure to give us better consideration for the natural environment, as well as helping us better understand its relationship with the whole.
- *Avoid totalizing tendencies.* As we survey relations in nature, we see that these are *boundless*. They are far more than we can understand or control. Therefore, we shall know better than to subject the environment increasingly to human control and management.[8]

4. Sautet, *Nietzsche for Beginners*.
5. Attenborough, "Prince William," 11:15.
6. Rolston, email to author, Nov. 17, 2019.
7. This contrasts with a common religious view: humans are the stewards of creation. Interestingly, the biblical account reads (abridged), "I have given you every plant yielding seed, and to every beast every green plant" (Gen 1:29–30).
8. Science journalist John Horgan noted, "Natural systems are open: our knowledge

- *Know how you are situated in the world.* If we have a good grasp of our own relationship with nature, as to how we influence nature and why—and how nature influences us—we shall nurture a balanced and sustainable relationship with nature.

Again, these maxims are based on relations between things. More than that, they are based on the *features* of relations as a whole—their *meta-features*: their totality and boundlessness, and our finitude.

Consider a single pond. It is related to the water cycle, food webs, nutrient levels, geological formations, salinity, shading, grazing, and uncountable things besides. It may even (credibly) be related to the stars, plankton in the sea, or a sunny day in Antarctica. Ultimately, it lies beyond our understanding.[9] Yet this is only one pond.

As we lift our eyes above the pond, we see hills and valleys beyond, flora and fauna, rivers and clouds—all of which are intricately interrelated.

If a single pond should lie beyond our understanding—even our *ability* to understand—how much more the entirety of nature.

Granted, we understand a lot about our role in nature's decline, and we have immensely powerful tools to study it—yet it is a mistake to assume that this is or ever will be enough to solve the problems of nature.

The question basically is this: do we *know* enough, do we *understand* enough, and do we have the *power* to rescue nature? The answer from experience is no, and the evidence is plain to see.

But a negative answer comes not only from experience. It is no *in principle*, too. Not only is our knowledge vastly inadequate to the task, but it is hopelessly *exclusionary*. It lies in the very nature of knowledge today to exclude the whole. Nor can we assume that, because we have been able to restore parts of nature in the past, we shall do so again in the future. Our situation now is new. In due course, we think on this in more detail.

Where then may a solution lie? Various base principles have been suggested our for dealing with nature:

- Nature is vital to human flourishing.
- Its health is essential for economic progress.

---

of them is always partial, approximate at best" (Horgan, "Complexity to Perplexity," 106). Science historian Naomi Oreskes wrote, "Verification and validation of numerical models of natural systems is impossible" (Oreskes, "Verification, Validiation," 641).

9. David Attenborough summed up our limitations: "We can exterminate whole ecosystems without even noticing it" (Attenborough, "Prince William," 0:30).

- The condition of nature is an indication of our social well-being.
- Every form of life has a special goal.
- Nature enjoys a *sacred* status.

Beyond all such considerations, however, it may be the very complexity of nature that promises an answer. This complexity helps us both to define and to address the epistemic crisis.

The natural environment presents us with boundless relations. The ecologist Kenneth Melanby spoke of an infinity of overlapping environments—each of which contains a vast complexity of its own.[10] The environment is the *ultimate zone*, in which a vast tissue of influences cannot be reduced. A reductionist approach does not apply.[11] Science writer Kitty Ferguson wrote, "A deeper understanding of nature . . . lies in recognizing that much of it will never reduce to simplicity and linearity."[12]

In fact, nature seems to stand as a vast and all-encompassing symbol to teach us that the project of reason is trammeled. Not only this. For want of a concept of how nature *ought* to be, we cannot say that it *ought* to be this way or that: it ought to be so and so constituted or so and so engineered.[13] Here, we can only surrender ourselves to what is.

> Our immediate environment is not nature as before, but organization. Yet this immunity from nature produces a new crop of dangers, which is the very organization.
>
> —DIETRICH BONHOEFFER

---

10. Melanby, "Environment," 275.

11. The philosopher-economist Friedrich von Hayek wrote, "There is danger in the exuberant feeling of ever growing power . . . to subject not only our natural but also our human environment to the control of a human will" (Von Hayek, "Pretence of Knowledge," last para.).

12. Ferguson, *Fire in the Equations*, 275.

13. This must mean that the environment lies beyond ethics. Pope Francis wrote that nature, rather than being "a system which can be studied, understood, and controlled . . . can only be understood as a gift from the outstretched hand of the Father of all" (Francis, "*Laudato Si'*," §73).

Entrenching Taboos

We should not underestimate the *invasiveness* of the built environment, which is the human sphere. Every time we put down a house, build a dam, or plant a field, we must push back nature.

Suppose that I have just visited the remote interior of Africa. It seems strange to me now, returning to the city, as I look down on human habitation from a mountain pass. All the jackals, porcupines, blister beetles, hover flies, kestrels, owls, which I saw in the interior, are wiped out across this vast expanse of the city. There is no place for them here. Wherever humanity spreads its habitation, they must be expelled.

Today, we know, too, that the built environment impacts nature not only within the built environment, but it deeply impacts the *natural* environment—even *protected* environments.[14] How, then, may we order our built environments, so that the natural environment may continue to thrive? It comes down, again, to reason's inadequacy to the task.

In times past, the natural environment was in many ways *taboo*. It was often forbidden territory (in Latin, *prohibitos autem terra*). It was *under the ban*. In some cases, it was more valuable than the life of one who plundered it or desecrated it. In some societies, the plunder or desecration of the environment was punishable by death: for example, harvesting out of season or fishing in an area demarcated for another clan.[15]

On the basis of a belief in the environment as *forbidden territory*, there is a principled way forward. The environment is not merely an object of wonder and awe, nor is it sentimentally sacred. It is out of bounds and set apart—as it were, *religiously*. Whatever the value of nature—whatever its stores, whatever its appeal, we must not touch it—not even

---

14. In my boyhood, in the 1960s, there was a massive die-off of birds in the Pacific. Worse, it happened in a wildlife refuge. Seventy-four percent of one hundred dead albatrosses had ingested 241 items of plastic (Kenyon and Kridler, "Laysan Albatrosses," 341).

15. I have this through conversation with village elders on Abaiang and Beru atolls in the Pacific. Martin Heidegger criticized the view that all of nature is a "standing reserve" (Heidegger, *Question Concerning Technology*, 17).

with the desire to regenerate it.[16] The only way forward is to treat it as off-limits and hands off, and this is nonnegotiable.[17]

> ### Excursus
>
> How, then, should it be possible for nature to be off-limits and hands off today? Is not humanity everywhere roaming beyond its bounds? Is it possible or practical, that our encroachment on nature should be stopped?
>
> The notion of such separation is not merely a wistful one. There are large areas today that we do keep off-limits: contaminated areas, military zones, and diamond areas, for instance. These are *as good as* taboo.[18] Yet by comparison, we treat the natural environment with ambivalence.
>
> Just one hundred years ago, nature did not need any help from humanity. That is, only in our recent past, we knew a world in which nature was allowed to be nature and did not need human cosseting or monitoring. The deliberate, nationwide, even global protection of the natural environment through human custodians is a recent development.

We must leave nature to nature. It will be restored only when the equilibrium of the natural environment is maintained in every part without human intervention—or, rather, when its *dynamic* equilibrium is maintained both globally and over all of its various cycles, as a *self-sustaining* equilibrium.[19]

16. The philosopher-journalist Andy Lamey observed, "Thinking of ecosystems as spontaneous orders generates a presumption against interfering with their natural functioning in a manner that results in anthropogenic species loss" (Lamey, "Ecosystems as Spontaneous Orders," 64).

17. By way of contrast, *deep ecology* holds that a holistic view of nature includes humanity ("Deep Ecology"). Although natural and built environments inevitably overlap, I do not consider this ultimately to be tenable. Our material culture is no longer in harmony with nature.

18. In 2020–2021, nature was separated in a different kind of way when, due to the COVID-19 pandemic, wildlife reclaimed the built environment in many places.

19. One may call this a climax, which is the climactic state of an undisturbed vegetational community (Allaby, "Climax Vegetation," in *Dictionary of Ecology*, 79).

The biologist Edward O. Wilson advocated the protection of half the land and sea on the planet, as a "first, emergency solution."[20] However, he assumed thereby that we have the ability, *over time*, to master the problem of nature's decline. On the basis of a philosophy of relations, nature is impossible to master. There must be a complete and permanent separation of the natural and built environments, as far as this is possible.[21]

Finally, in terms of relations, there is something to be said for the view that many, if not all forms of life are vital to every other. They live not in and for themselves,[22] but every life form, whether it be an elephant, a midge, or a fern, is intimately involved with every other and should be regarded as important to the whole.[23]

Just as our ability to think expansively and integratively depends on the availability of all information to our minds, so an expansive and integrated environment depends on all parts being present and available to its totality.[24] "Each ecosystem," wrote Wilson, "is a web of specialized organisms braided and woven together."[25]

---

20. E. Wilson, *Half-Earth*, prologue.

21. Humans cannot be *completely* separated from nature. Humans breathe oxygen, oxygen is generated by photosynthesis, and so on.

22. This will have a lot to do with how *robust* we consider nature to be—that is, how important various organisms are to the whole. Religious views tend to believe that every organism is essential to the whole.

23. There is now the debate, which we could not have imagined even recently, as to *which* species we should save and which we should let go (Brandl, "Which Species?"). In principle, this is not a decision we are capable of making.

24. For this reason, it is questionable whether it will be possible to colonize planets other than Earth, sustainably.

25. E. Wilson, *Half-Earth*, ch. 10.

Chapter 21

Unvirtuous Behavior

*Unvirtuous behavior, in all its forms, has to do with
one's conceptual arrangement of the world.*

As we approach the end of our tour of ethics, both personal and public, we have dealt mostly with *virtue*, or principles of moral excellence. Yet while philosophy tends to devote more attention to virtue than vice, in real life, futile, hurtful, and harmful behavior looms large.

Not only do we find foolishness and fallacy in every sphere. We find bluff, deceit, rationalization, subterfuge, lies, and half-truths—to name just some of the many forms of unvirtuous behavior we encounter almost daily.

Then, too, we find *violence* in many forms: physical, emotional, verbal, financial, sexual, psychical, systemic, and structural. This, too, we intuitively feel, has everything to do with a loss of virtue or something of the sort.[1]

There are various theories as to why one's behavior becomes unvirtuous. One of the more familiar theories is Aristotle's. Departures from the *golden mean*, Aristotle wrote—which is departures from the middle way between extremes—may lead us to excess on the one hand, deficiency on the other.

---

1. While we do not attempt here to explain the origin of moral and physical evil, we seek to identify a dynamic that is common to all unvirtuous behavior.

> **Excursus**
>
> However, we soon encounter a question of *definition*. What do we call "un-virtue"? What are these behaviors, as a *class*? Is there a name for a class that unites all things that lack virtue?
>
> We may try various terms to subsume them all: falsehood, dysfunction, untruth, and so on. But in some way or other, all these terms seem inadequate or fail at some point.
>
> "Unvirtuous behavior" seems to be the most promising. Further, it is associated with various oppositions of virtue: for example, behavior that lacks dignity; wants merit; is unjust, unbalanced, or dishonorable.

Aristotle suggested *courage* as an example.[2] Too *much* courage becomes recklessness, while too *little* becomes cowardice. This view lies not far from the argument that is offered here.[3] When unvirtuous behavior overcomes us, this has to do with a view of the world that is not balanced and broad.

It would help us first to understand the wider context in which all of our behavior occurs, then to set our negative behaviors in this context.

Suppose that the image I have of my life is one of a happy family in suburbia—a friendly dog, fresh muffins on the table, and daisy chains and laughs. I look from my kitchen window, suddenly to see my little girl with her face down in the grass. There is a disjoint—a conflict with the picture that I have in my mind—and I spring into action. We are, after all, motivated by the unexpected.

Therefore, prior to all of my actions—and *driving* my actions in every case—is the way in which I arrange the world in my mind.[4]

---

2. M. Beard, "Courage."

3. Aristotle did not attribute *all* unvirtuous behavior to a departure from the golden mean. Some attitudes and actions were wrong in themselves, among them envy, murder, and theft (Kraut, "Aristotle's Ethics," §5.2).

4. Jean-Paul Sartre held that the goal is the coincidence of the subject's consciousness with its own subjective reality. This is (by his particular definition) *authenticity* (Golomb, "Sartre's Early Phenomenology," 335).

Different people have different conceptions of the world and so will spring into action for different reasons. Some may not *want* a happy family in suburbia, a dog, and fresh muffins on the table. Some may want to immerse themselves in figures. Some may want to be loose and wild. Some may want to live on Mars. The possibilities are as many as the people.

Suppose I am an aging mobster. The image that I have of my life is one of loyalty to a tightly knit group, under the absolute authority of its elders. I open the newspaper one morning, suddenly to see my picture on page 3, where I am accused of racketeering. A nephew has spilled the beans. Again there is a disjoint—a conflict with the picture that I have in my mind. "My nephew has ratted on us!" I cry. "He will pay dearly for this!" and I spring into action.

Forms of Unvirtuous Behavior

In chapter 10, we described this dynamic—too simply, we said—in terms of the principle of contradiction. Our view of the world does not *match* the world, and on this basis, we act. Now in the context of unvirtuous behavior, we are in a position to study some subtleties and complexities of this dynamic.

Casting our minds back to earlier chapters, in terms of the relations that we trace, our conception of the world may be balanced and broad, or parochial, short-sighted, and self-interested. Some will live a "large" life—characterized by expansive and holistic thinking—well-rounded and meaningful; while others will live a small-time existence, as fools or bunglers. In short, some will become wise, and some will become fools.

In its most innocent form, unvirtuous behavior is rooted in the "small" view of life, which is characterized by mundane and insular thoughts. Wherever we find it, we tend to pity it, laugh at it, or denigrate it, but we don't much take it to heart. It matters little to the rest of the world whether someone builds a house on sand, tries to cross the sea in a bathtub, or measures his social standing by the number of drinks he can down at the pub.

Yet the small view need not be quite that absurd. It may include all who devote the best hours of their lives to lesser ends: say, the cleanliness of a home, the well-being of a cat, a fixation with the girl in the choir—and in many cases, their own personal gain. This is unvirtuous behavior for the reason that, in itself, it does not evidence a conception of the world that is balanced and broad.

With these simple observations, we may now describe the first of three forms of unvirtuous behavior—call it *foolishness*. Foolishness misses the fundamental ethical maxim:

- *Think expansively and holistically.*

We find, then, a *second* form of unvirtuous behavior, namely, lies and deceit.

Again, we all arrange our worlds—which we *hold up* to the world—in different ways in our minds. Again, from these arrangements, our motivations arise. Yet given that we each have different conceptions of the world, we each have different motivations. It stands to reason, then, that my own motivations may come into conflict with the motivations of others—and if I do not yield to them, then they must yield to me.

If, therefore, another person cannot change my conceptual arrangement of the world through natural influences or inducements, they may yet be able to change the conceptual stuff of which my arrangements are made. With a few targeted ruses, they may change the world I *think* that I inhabit.

Consider an imaginary scenario. I feel passionate about the village pond, while another man wants to build a parking lot over it. If he cannot overcome my passion for the pond, by seeking in good faith to change my conceptual arrangement of the world, he may tamper instead with the conceptual stuff with which I work.

He may persuade me (falsely) that permission for his car park has been granted on high authority, that the pond is doomed by a falling water table, or that all ponds are death traps for children. This now differs from mere foolishness, in that it seeks to manipulate what I know. It changes the *things* in my life, as I imagine them.[5]

This happens all the time—among other things, on the personal level of lies, on the corporate level of misinformation, and on the political level of propaganda—a prime example being Plato's "noble lie."[6] Such behaviors fail the test. Not only do they reveal a conception of the world

---

5. Immanuel Kant might have called such things *phenomena*, as opposed to *noumena*, if phenomena include both concrete objects and abstracta.

6. Plato, *Republic*, 301. While this refers to a *leader's* "noble lie," the question also arises as to whether a *citizen* may tell a noble lie—say, to save an innocent person's life. Presumably, yes, where this is born of expansive and holistic thinking.

that is not balanced and broad, but they deliberately trace false relations, or reduced or distorted relations, in this world.

This, too, misses one of the fundamental maxims of ethics:

- *Embrace those relations that exist in the world.*

Excursus

Someone might counter, "Not *every* false conceptual arrangement of the world may be judged as *morally* wrong. Take false *theories*, for instance, or false religious beliefs—even the false arrangement of my dinner table."

Yet how should we say that such things are not *morally* wrong? On what basis may we call them, say, cultural norms or common practice instead? There is no rational basis on which we may separate them from moral—or immoral—behavior.[7]

Therefore, our concept of that which counts as *moral* may need to change. We first raised this in chapter 13, in the context of esthetics, where we said that esthetics and ethics broadly overlap.

We may say much the same about a yet more serious form of unvirtuous behavior. Not only may another person seek to change the way in which I arrange the world in my mind. They may change the world itself—through force, violence, or actions of the sort.

Consider again the man who wishes to build a car park over the village pond. In the dark of night now, he sends a small-time crook with a dump truck to fill in the pond in one dramatic act. By the morning, the pond is gone. Now my conceptual arrangement of the world must change, because the world itself has changed.[8]

The dynamics, in reality, may be more complex than this. It may be easy to see that a pond was filled in on the orders of one who had a vested interest in it. It may be less easy to see that running me out of town

---

7. In the theological realm, this may greatly increase the thoughts and acts that are included under sin.

8. But consider also the possibility that one obstructs or disrupts harmful behavior. Non-violent direct action (NVDA) is a powerful strategy.

with false rumors had to do with the pond or that someone now drives a new luxury car on this account. And so the world of unvirtuous behavior becomes tangled and dark—with deceit, threats, and violence as vast as the ocean.

Let us now notice what has happened in the course of this short chapter. By means of some basic principles, we have reconciled all manner of evils. Whether someone is reckoned a fool, a liar, or a thug, these are basically one and the same. They all relate to the ways in which we go about with things and relations. Regardless of the kind of unvirtuous behavior to which we succumb, it happens through a false conceptual arrangement of the world.

Now we may notice something important about human nature. We have seen that our moral integrity, or lack of it, is closely related to our worldview—which is how we arrange the world in our minds.

In many cases, unvirtuous behavior changes only when our worldview changes, through a rearrangement of the relations that we trace in this world.[9] Unvirtuous behavior first has to do not with greed or with need, with compulsion or coercion, but with first philosophies.

Abnormal Psychology

Unvirtuous behavior has a lot to do with *psychology*: in particular, abnormal psychology, which studies dysfunctional patterns of thought, emotion, and behavior. Historically, the origins of dysfunctional behavior have been interpreted in various ways. There are religious, biological, and psychological explanations for mental or emotional dysfunction, as well as theories of multiple causality.

> The solution [to pathology] means the setting forth of all the typical patterns or modes of arrangement into which mental processes fall.
>
> —MADISON BENTLEY

---

9. This is, needless to say, a slow process. Our understanding of the world changes slowly, and so, too, do our moral predilections.

In terms of this philosophy, all motivations and, therefore, all behaviors have their origin in the way that we arrange the world in our minds. Therefore, while there is no doubt that we are physically predisposed towards certain behaviors—through genetic factors, illness, weariness, medication, or the pressures of the moment, among other things—there can be (almost) no behavior at all without the mental models that drive it.

In terms of this metaphysics, then, the preferred approach to treating mental disorders is to help people see how their behavior relates to the *whole*: to understand how they act upon the world and how the world acts upon them—in relation to their own body and mind; their personal and social relationships; their relationship to ideology and religion; and their relationship to the past, present, and future.

This means that the best counselor or psychologist will be in a position to help others see themselves through different eyes—eyes which already see the world in broad and balanced relation. The primary goal, then—without necessarily discarding other treatments for mental or emotional dysfunction—is to obtain a healthy, expansive view of the world.

Chapter 22

Critical Theory

Critical theory has become increasingly important in our time— yet its core characteristics are deeper and broader than we have supposed.

CRITICAL THEORY, IN THE words of the philosopher Max Horkheimer, seeks "human emancipation."[1] It is, wrote philosophy professor James Bohman, "social inquiry aimed at decreasing domination and increasing freedom."[2]

The theory is "critical" because it is not neutral—which is to say, it is *normative*. Professor of communication and culture Robert M. Seiler of the University of Calgary wrote, "Criticism involves . . . judgments for the purpose of bringing about positive change."[3]

In an important sense, critical theory goes back to the ancients. Aristotle, we said, thought of ethics as the golden mean, or the balanced life. And Socrates, in defining virtue, said, "Choose the mean, and avoid the extremes on either side, as far as possible."[4]

Sound ethics, therefore, represents the achievement of *balance*, both in the human person and in society: balance between unity and diversity, novelty and tradition, thought and feeling, individual and community, and

---

1. Bohman, "Critical Theory," §2.
2. Bohman, "Critical Theory," §1.
3. Seiler, "Human Communication," §1.
4. Plato, *Republic*, 301.

so much more. It is not hard to see how this coincides with critical theory, which seeks to bring balance to social inequalities of various kinds.

Critical Knowledge

Now it stands to reason that, as we seek to balance all things, we have *knowledge* of those things. We are informed of them before we begin to balance them. In fact, if information is lacking as we make judgments about society, our resulting balance must be askew. This ties in with chapter 15, where we noted the critical importance of information to political systems.

In order to control people, one simply adds or removes information—which is the knowledge of relations between things—to pitch the balance in one's favor. Worse, one may *suppress* such information through violence, compulsion, intimidation, and various other means, eliminating unwanted individuals or seizing control of systems, to neutralize such information as is not wanted.

In the previous chapter, we said that unvirtuous behavior is in large part about people who change the world that I *think* I inhabit—alternatively, who change the world itself—through various coercive means. Therefore, oppression may be defined as a loss of information to the system, through people's actions—sometimes unthinking but often intentional.[5]

Oppression, therefore, rests on the suppression of information. Alternatively, it rests on a failure of the absorption of information. One may have systems that seem perfectly friendly towards all information yet, in practice, fail to act on it. Information itself goes hand in hand with its reception, incorporation, and, of course, pursuant action.

---

[5]. Arguably, not all obstruction or corruption of information will qualify as oppression. It may be due to human limitations or failures, for instance.

Excursus

Yet may oppression really be reduced to information bottlenecks? In a great many cases, yes. Wherever we look at history, we find that the suppression of information has lain at the root of oppression.[6] For example:

- The Dreyfus Affair, where high-ranking military officials suppressed the report of Marie-Georges Picquart.[7]
- The destruction of Bikini Atoll, where "information was withheld," from islanders in particular.[8]
- The murder of George Floyd, where it was only through a fortuitous video that police "information" was contradicted.[9]

With the central role that information plays in oppression, one may, therefore, *identify* oppression by examining where information is suppressed. Further, one has a method through which one can *remove* oppression, which is the removal of the suppression of information or the lack of its uptake.

This has three implications for critical theory.

- Firstly, where information is suppressed, this may not in every case happen along recognized class lines—or lines of race, gender, privilege, and so on. This is a deficiency of critical theory today. Information is suppressed along all kinds of lines, which critical theory may fail to identify. It may miss oppression that we had not imagined or that lies beyond our familiar categories. This should not be understood as a rejection of critical theory. Rather, critical theory as we know it does not drive deep enough.

---

6. Various *kinds* of information may be important. Apart from cognitive information, affective information is particularly important, having to do with emotion and perception.
7. Read, *Dreyfus Affair*, 175.
8. Faleomavaega and Flake, "Commending the People," recital 9.
9. Levenson, "What We Know Now," para. 6.

- Secondly, when one speaks in terms of the suppression of information, one broadens the scope of critical theory. Among other things, one may now speak of the oppression of the environment.[10] Where, through the suppression of information, we fail to include the environment in our thinking, we oppress wetlands, insects, elephants, forests, fish, and so much more. Critical theory is not fully equal to such forms of oppression, precisely at a time where they threaten the ruination of our world.

- Thirdly, critical theory has often been associated with "cancel culture." The purpose here is not to discuss the merits or demerits of cancel culture but to note that one should take care that cancel culture does not limit the freedom of information. The loss of such information to the system could signal oppression.

The Heart of Oppression

Let us notice now that, in terms of philosophical categories, this chapter represents another example of low-level thinking. It goes down to bedrock. It goes down again to the things-relations distinction, which originated with the ancient Greeks. It is when the relation of things to things is distorted or rejected that oppression occurs.

Oppression may now be defined as a loss of information to the system through various kinds of pressure, including physical coercion.[11] In short, this describes critical theory, which exposes oppression—yet more than critical theory, it goes to the very heart of reality, which is the relatedness of all things. It goes beyond *human* oppression, too, and includes our long and sorry oppression of the environment, where we failed to take into account all the information we should have done. One may call it *new* critical theory.

New critical theory has important advantages. It shifts the emphasis from issues of oppression to issues of information. Thus it lowers the temperature, broadens the theory's purview, brings oppression into sharper focus, and offers a specific means of dismantling domination.

---

10. While critical theory does address issues of the environment, this tends to be in the context of social transformation: "Environmental issues evoke questions about relations between and among ourselves" (Dyer, "Introducing Green Theory," para. 2).

11. We may further oppress our own selves, by failing to take into account the information that is required for a balanced existence.

Chapter 23

Awakening

As much as we have said about ethics so far, we cannot claim to have established more than a *voluntary* ethics. An awakening is required.

THE MOST IMPORTANT THING that we have said about a sound ethics so far is that we should think expansively and holistically—relating things to things in a wide-ranging way—to cultivate a conception of the world that is balanced and broad. Not all people think in this way. In the previous chapter, we said that people's thinking may, on the contrary, be parochial, short-sighted, or self-interested.

This immediately goes to something deeper. Not all people are open to the way that relations reveal themselves to us in their totality—and some are more open than others.[1] That is to say, while we shall not be able to survey every relation in an endless expanse, we shall in principle be open to them all.

In order to advance on the path of progress, we need first to reach a fundamental *decision*. We must be willing to *abandon everything*—culture, wealth, prestige, comfort, even truth—as we conceive of truth—in order to arrive at truth.[2] If we value anything more highly—even a single

---

1. The Muslim scholar Jafar al-Sadiq held that the "fortieth Greater Sin" of Islam is to commit a lesser sin repeatedly. "Consistency upon the smaller sins" constituted a greater sin, thus pointing to an inner orientation (Shirazi, "Fortieth Greater Sin").

2. In a sense, this is not a search for truth. It is the *loss* of truth. As long as we cling to particular ideas of truth, we may fail to make the progress we need.

thing—this will tend to restrict, confine, reduce, and distort the truth through our own prejudices and preconceptions. We must be willing to abandon, if need be, all the relations we have ever traced, even if this should cost us everything.[3]

Ultimately, this means that our behavior is an issue of *orientation*. Rather than being about distinct acts that we can name and define and categorize, it is about something on a far more basic level:[4] our acceptance—or not—of the relatedness of all things. Here again, we find that a low-level idea is a core determinant of life as we know it: the recognition of a boundless expanse of relations, no more and no less.

> Reality's ultimate reason, support, goal, its primal source, primal meaning and primal value are laid open as soon as he [that is, we] lays himself open.
>
> —HANS KÜNG

There is another way of putting this. In earlier chapters, we saw that there are two fundamental modes of thinking in this world. Either we are held captive by an unacknowledged metaphysics, or we take the bird's-eye view, to survey the wider features of relations—*meta-features*—from the undisturbed vantage point of a distance. We no longer deal with relations between things in their *immediacy* but survey their features as a whole.

So different are these two modes of thought that we may consider the switch from the one to the other to be an *awakening*.[5]

---

3. This implies that awakening may come through suffering, pain, or despair. Some might wish to avoid this.

4. Applied to religion, this differs markedly from Immanuel Kant's position: "Apart from a good way of life, anything further which the human being supposes that he can do to become pleasing to God is a mere religious delusion and a pseudoservice of God" (Kant, *Religion within the Limits*, 189).

5. This differs from the recent term *wokeism*, in that an awakening does not make one fully awake. One has only just awoken. In a sense, the term *awakening* is more radical than wokeism.

Dead Ends

It would be helpful to put our thinking into reverse. Consider what happens when we *leave* the bird's-eye view—from which we see emergent features of relations between things—to enter again into a self-contained, self-regulating system—to become bound once more by the rules of linguistic systems and an unacknowledged metaphysics.

We now no longer take the broad view. We gravitate once more towards centers, starting points, basic principles, and the like, once more putting dominoes end to end, as it were. Again, to use Derrida's term, we return to *origins*—or, to say much the same thing, we *launch forth* from origins.

This *manifests* itself in ideologies, ideals, worldviews, narratives—one thinks again of the "single thought"—all of which are *partial*. Yet because they are partial, our thinking becomes vulnerable to all the ethical weaknesses we have surveyed: among them, a parochial view of the world, totalizing tendencies, and the creation of arbitrary centers.

But this has real consequences. With hobbled worldviews, we inevitably come up against personal, social, political, and environmental *dead ends*—and these dead ends may cause intellectual, emotional, and physical crises.

Historically, the feudal system was a dead end, Nazism was a dead end, Soviet socialism was a dead end—and modernity, too, now looks like a dead end. On a smaller scale, there have been dust bowls, food queues, urban decay, and hyperinflation—all dead ends in their own more limited contexts. And on a most personal level, we are all familiar with dead ends: burnout, addiction, divorce, and heart attack—above all, emotional and (some would say) *spiritual* crisis—which threaten our ability to function.

There comes a point, then, where dead ends become *untenable*.[6] People begin to feel it and know it. This may be recognized intellectually or through one's whole being—namely, that one exists in the wrong *mode* of thought.

Imagine that our first philosophies are something like a painting on a wall. We trace the texture of its paint with our finger and follow its contours with our eye—yet we understand little about the painting until we step

---

6. Rabindranath Tagore observed, "Sufferings free the mind from its 'habit-environment'" (Tagore, *English Writings*, 44). Karl Jaspers held that we come to know ourselves most vividly in "limit situations" (Thornhill, "Karl Jaspers," §4).

back and view it as a whole. It is then that we see its meaning. So it is, too, with awakening. We step back from a bounded, limited view of the world, where secondary pursuits overshadow primary pursuits, and we see things in a new way. We see the broader features of relations and reality.

Excursus

What, then, of people who are in a dead end but do not *know* it? Should we live and let live? Or should we draw their *attention* to their plight[7]—engaging with them, confronting them?

We may think on some of the examples above: Nazism, for instance, or modernity—and on a simpler level, food queues, or burnout. Astute observers may *predict* that people, in such situations, are entering dead ends.

If, as we have said, ethics is intimately bound up with compassion, then it is a *compassionate* act to bring people to a recognition of dead ends. This goes to the question of awakening, too, since, if there is no *recognition* of dead ends, there may be no awakening.

Corollaries

An awakening not only changes our thinking but changes the way that we live. An awakening has to do with the relations that we trace between things—and the relations that we trace between things are the foundation of our character and behavior[8]—in fact, the foundation of everything we see in this world.

We now make an unexpected observation. We find that we have, at a stroke, divided all of humanity into two types of people: the awakened

---

7. Francis Schaeffer held that one should drive people to recognize their dead ends, even at the risk of their suicide (Schaeffer, *God Who Is There*, 159).

8. We may notice that one who is awakened not only *behaves* differently; she *influences* differently. As she seeks to banish from her life the old rational principles that enslave—whether political ideologies, social mores, or personal repression—she becomes a beacon in her world. Only she who has grown from the depth of her inner self can rise above the medium of her society, to be a real communicator.

and the unawakened. This need not be visible or easily discernible, yet it may fundamentally influence people's ethics and actions.

To recognize the difference, we need a discernment that penetrates appearances. It is wisdom to understand that one lives in a world of wheat and of tares, of sheep and of goats.

More than this: wherever one finds deceivers, backstabbers, wolves in sheep's clothing, whitewashed tombs and all, it is due to this division that runs through all of humanity. It is the *superficial* look that sees a world full of *deeds*, good and bad. It is the profound look that sees the awakened and the unawakened.

While this is language that one often finds in "religious speak," particularly in connection with *conversion*, we need to note that the religious person may not be awakened as here described. We are speaking about the abandonment of everything we may use to justify our existence and the disintegration of our first philosophies,[9] while religious conversion may merely modify our first philosophies or exchange them for another set of analogous beliefs.

---

9. It is a form of nihilism. To borrow words from the philosophical theologian Kevin Hart, it "requires us to revalue all our values" (Hart, *Postmdernism*, 34). At the same time, note that "the abandonment of everything" implies the absolute priority of what is new.

Part IV

Science and Math

The fourth part of this book addresses the crucial fact-value distinction, first with the purpose of reconciling fact and value. It continues then by sketching what such reconciliation signifies in science and mathematics above all.

Chapter 24

Defining Ought

The meaning of the word "ought" is much broader than that of moral obligation. Importantly, it encompasses science and mathematics.

THE MEANING OF THE word "ought" is crucial to the coming chapters. It is the one word we need to understand in order to reconcile our natural language—filled with feelings, judgments, opinions, commands—with science and mathematics in particular. In fact, it is crucial to the creation of an integrated metaphysics.

This single word has addled our thinking in crucial areas and has often blocked the path to progress. It serves as another evidence of basic concepts that infiltrate our thinking everywhere, with potent effect.

The word "ought" lies at the heart of the *fact-value distinction*, which is basic to the problem of a rational ethics—but, more than that, it is basic to the separation of all things that *are* from all things that *ought to be*. Among other things, "ought" lies at the core of the separation of mathematics and science from much of our everyday thought and language. Hard science in particular is said to represent things that *are*, while our ordinary daily discourse so often enters upon the way things ought to be.[1]

But the meaning of the word "ought" applies not only to our everyday experience. We shall see that much of the language of math and

---

1. This is merely the prelude to a full solution. Once we have taken "ought" out of the problem, we find that we are left with a problem of different *classes* of things: scientific, mathematical, and everyday things. We address this later in pt. 4.

science, which we intuitively think of as "is," really represents "ought." This has an important corollary: the certain sciences are uncertain, objective statements are subjective, and facts are mere points of view. This puts us in a dangerous place. But we are running ahead of ourselves.

On first impression, the meaning of "ought" might seem clear. Yet the popular notion of the word—namely, that it refers to things we *ought* to be doing—does not stand up to scrutiny. To put it another way, the concept *behind* the word is muddled.

In order to discuss the unification of the various "languages" we speak—which we briefly introduced in chapter 2—above all, ordinary language, scientific language, and mathematical language—it is critically important that we first clear up any possible confusion surrounding this word.

There is a popular view—and quite a strong one at that—that "ought" applies only to *moral obligation*. For example, "You are behaving badly. You ought to know better," or "You owe me the money. You ought to pay it." Ought, from this point of view, is a *moral* ought. Yet not all oughts are moral, by far, as we shall see.

Importantly, if one holds that the word "ought" applies only to moral obligation, it follows that it does *not* apply to *science*. Philosophy professor James Ladyman summed up a widely accepted view like this: "Science abolishes the person."[2] It is impartial. It has nothing to do with human judgment—at least, not in the sense of being mixed up with the subjective world of human values.

Meanings of Ought

"Ought" has many meanings. Its meaning varies with context, and sometimes, it may be ambiguous. So great is the variation that we can advance merely a few examples.[3]

- Apart from indicating moral obligation, "ought" is often used to indicate that something is *probable*: "They ought to be there by now," or "Ten minutes ought to be time enough."

---

2. Ladyman, in Bragg, "Human Nature," 31:02. This is not Ladyman's own view.

3. The philosopher and economist John Broome differentiated between normative and nonnormative oughts, owned and unowned oughts, qualified and unqualified oughts, and objective and prospective oughts (Broome, *Rationality through Reasoning*, ch. 2).

- There are those "oughts," too, that have to do with something that is advisable or desirable: "We ought to take him a gift," or "You ought to try the sushi."

- Ought may apply equally to *material* things: "This system ought to work," or "You ought to get a deflection of 1.75 arcseconds" (the famous Eddington experiment).[4]

We further find various *synonyms* of ought: above all, "must" and "should." Yet while there has been a great deal of debate about these words and various attempts to separate them, their meanings overlap greatly. Catherine Soanes, a project editor for *Oxford English Dictionaries*, observed, "We can use must, should, and ought to express broadly similar meanings: the main distinctions between them are related to degrees of emphasis."[5]

Expectation

Is there anything, then, that unites all of our uses of "ought"—perhaps, too, its various synonyms? In fact, there is. Almost all are about *expectation*. Not expectation in the sense of *anticipation* but in the sense that one believes or feels that something will or should happen.

"I *expect* that you will behave like this," "I *expect* that they will be there by now," "I *expect* that we shall take him a gift," "I *expect* that you will get a deflection of 1.75 arcseconds"—and so on. Sometimes such expectation may be strong, sometimes weak, and sometimes it may appear in disguise—yet mostly, it is present where the word "ought" is present.[6]

In fact, the kinds of words that have something to do with expectation are legion. We find them everywhere in our language—in ordinary sentences like "This will work" (I expect it to work), "Let him go" (I expect you to let him go), "You have to do it" (I expect you to do it), and so on—a kind of thinking that is fundamental to our existence, in that we find it all over.

---

4. The experiment of May 29, 1919, to measure the deflection of starlight during a total solar eclipse. A *Merriam-Webster Dictionary* example: "You ought to get nine for the answer."

5. Soanes, "Must, Should, or Ought," §1.

6. One may use "ought," too, of things one does *not* expect to happen. For example, "He ought to have a medal" or "There ought to be two moons in the night sky." There is, however, still a figurative relation to expectation in such cases.

> Experience has shown that, hitherto, the frequent repetition of some uniform succession or coexistence has been a cause of our expecting the same succession or coexistence on the next occasion.
>
> —BERTRAND RUSSELL

Not all of our uses of the word "ought"—not all, by far—are *moral* ones. Nor is this the case with its various synonyms. Just as we apply these words to human *behavior*, so we apply them to nonhuman things and situations.

It will be particularly important to us in the coming chapters that *scientists* use the word "ought"—together with various other words that express expectation.[7] A mere cursory examination of the major works of science reveals that they are replete with "ought." If they do not state what *ought* to be the case, they state what others *imagined* ought to be the case:

- "Chord AC ought to be greater than half of chord AD" (Nicolaus Copernicus).[8]
- "The same force ought to generate the same velocity" (Isaac Newton).[9]
- "We ought to find . . . closely allied forms" (Charles Darwin).[10]
- "We ought to obtain for the orbit of the planet an ellipse" (Albert Einstein).[11]
- "It ought to be possible to increase the magnetization" (Erwin Schrödinger).[12]
- "We can understand . . . more or less how things ought to behave" (Richard Feynman).[13]

---

7. Stephen Toulmin wrote about expectation in science, "The scientist's prior expectations are governed by certain rational ideas or conceptions of the regular order of Nature" (quoted in Pitt, *Theories of Explanation*, 195).

8. Copernicus, *Revolutions of Heavenly Spheres*, 35.

9. Newton, *Principia*, 735.

10. Darwin, *Origin of Species*, 354.

11. Quoted in Hiton, *Theory of Relativity*, 84.

12. Schrödinger, *What Is Life?*, 12.

13. Feynman, *Lectures on Physics*, 1:42:1.

- "A hot body ought to give off electromagnetic waves" (Stephen Hawking).[14]

In terms of this book, therefore, "ought" represents the way in which things ought to be related to one another, whether philosophically, theologically, scientifically, ethically, esthetically, or in any other field. John Broome made a major study of the word "ought." He wrote, "'Ought' is certainly not particularly a moral word."[15]

It is broader than that. It is important to remember this as we enter the following chapters.

---

14. Hawking, *Brief History of Time*, ch. 4.
15. Broome, *Rationality through Reasoning*, ch. 2.

Chapter 25

Fact and Value

The fact-value distinction is one of the biggest obstacles to a unified philosophy. We first treat this subject more generally, then in relation to science and math in particular.

SO MANY THINGS ARE done today because we *can*—where, previously, we could not.[1] We often choose to change the temperature, light, and sounds that surround us, where, previously, we could not; we censor millions of messages on the turn, which was impossible one generation ago; we order any which goods from any which place on the planet; we even alter animal and plant biology—including our own.

With the many things that we *can* do, which in the past we could not, what is there left that we ought or ought not to do? On what basis shall we decide, if it is not dictated to us by human will or necessity?

We first introduced the problem of "is" and "ought" in chapter 7. David Hume gave the problem its classical formulation: it is impossible to derive an ought from an is. That is, it is impossible to establish any value amidst an ocean of facts.[2] Value, it is said, is not found in the world.[3]

---

1. Much of the shift that we have seen in values has to do with our emancipation from *necessary* behaviors.

2. I primarily address David Hume's *is-ought* distinction. My discussion, however, applies equally to the fact-value distinction, which may or may not be seen as distinct from is-ought.

3. McDowell, "Aesthetic Value, Objectivity."

On the surface of it, Hume would seem to be unimpeachably right. The facts cannot tell us what to do. A statement of fact (for example, "We have the bomb") does not lead to a statement of value ("We ought to *drop* it").[4] Nor can the description of a mere phenomenon ("Many are chaste") lead to a statement of value ("They *ought* to be").[5]

The fact-value distinction has been with us for centuries. For centuries, too, philosophers have had the feeling that it ought not to exist. They have been strenuous in their efforts to set it aside—and yet it lives on.

Logic too is dumb. It gives no preference
to moral dispositions.

—HANS KÜNG

Things and Relations

The fact-value distinction goes to something deeper than fact and value. It goes to the things-relations distinction. In fact, Hume himself saw the importance of things and relations: "The distinction of vice and virtue is not founded merely on the relations of objects."[6] That is to say, we do not, by merely observing relations between objects, discover how they *ought* to be related.

Many philosophers have recognized the things-relations distinction as being central to philosophy. Russell wrote that many philosophers, following Kant, have maintained that relations are the work of the mind, while things in themselves have no relations. Wittgenstein, in his later philosophy, used a constellation of terms to refer to our reality as "things," "objects," "elements," "parts," and so on, while he thought that such things were combined as "arrangements" of things.[7]

---

4. Any field where "good" and "ought" apply falls within the range of axiology.

5. The sociologist and historian Max Weber pointed out that the mere fact that a phenomenon exists cannot establish its value (Blum, "Max Weber's Postulate," 46).

6. Hume, *Enquiry*, 172.

7. In spite of his reputation for innovation, in this area, Ludwig Wittgenstein was still quite old-fashioned.

One finds a similar view in the theory of *language*—or, rather, in the *cluster* of theories we have called structural linguistics. The philosopher Rudolph Carnap considered that, in language, *objects and their relations* are the topic.[8]

A pebble is a thing. A house is a thing. Even gravity, ideology, taxonomy are things (in chapter 3, we called them *abstracta*, which we equated with things). These things in turn may be related to other things. In a sense, even a *unicorn* is a thing, although we are unlikely ever to find one.

Things, then, may further be involved in what we call *truth conditions*—which means that they may be inserted into statements that can be affirmed or denied. When we affirm such statements, we call them "facts." For example, we insert the thing "pebble" into a statement ("A pebble sinks"), or we insert the thing "unicorn" ("The Scots keep unicorns"). Our things are now involved in truth conditions—which means that our world is filled with facts—and if not facts, then falsehoods.

Now that we have said a few things about facts and the things of which they are made, we may relate them to the fact-value distinction.

Value and Relations

Imagine a room that is empty—except for a four-legged table, which stands in the middle. Let us focus on the *relations* that we find there—at least, *some* of them. As we survey this simple scene, we notice that the table stands at such and such an angle to the walls and at such and such a distance from each one.

In terms of fact and value, we can now make two kinds of statement about the table. We may say, "The table *is* situated in the middle of the room," which is a statement of fact. Or we may say, "The table *ought* to be situated in the middle of the room"—or anywhere else for that matter—which is an expression of value.

Now notice that, in both cases, we are, in a sense, speaking about the same thing—which is the relative position of the table in the room.

Next, notice that it is possible for us to eliminate fact. In the first case, the table is where we think it ought to be. In the second case, the table is where we think it ought *not* to be—alternatively, is not where we think it *ought* to be.[9] The moral philosopher Philippa Foot recognized

---

8. Blackburn, "Formal/Material Mode," in *ODP*, 143.

9. Someone might object. If "the president is an autocrat," that cannot be construed

this strange interrelatedness of is and ought when she referred to "an oak that is as an oak should be."[10]

> ### Excursus
>
> Someone might object. If the table "is" in the middle of the room, this surely does not mean that it "ought" to be. One cannot play such tricks with words!
>
> Yet many "is" sentences imply an "ought." One may discover this whenever one *changes* a situation so as to change its "is." For instance, we load the table onto a dump truck, or we jam it in a doorway. Someone is bound to say, "It *ought* not to be on the dump truck!" or "It *ought* not to be in the doorway!" It ought to be where it was. It all had something to do with "ought" after all.
>
> Certainly, with these examples, we have imported new *contexts*. We have introduced a dump truck, and a doorway. Yet this shows us, too, that we thought the original *context* to be the right one.

In fact, we are surrounded by many examples, far simpler than this, of "ises" that are "oughts."[11] "My pen is not on my desk!" (it ought to be). "There is nobody here" (there ought to be somebody, anybody here). "What is this in my soup?" (it ought not to be there).[12] The "is," in all such cases and many more, means that it ought to be or ought not to be. In fact, the greater my certainty that something "is," the greater my sense of "ought," when I find it is not.[13]

---

as "the president *ought* to be an autocrat." It depends, however, how one is situated. Perhaps he ought to be an autocrat, perhaps he ought *not* to be; but in both cases, it is a matter of "ought."

10. Foot, *Natural Goodness*, 46. The philosopher Rudolf H. Lotze sought "the ground of that which is in that which ought to be" (Steinkraus and Mitias, *Taking Religious Claims Seriously*, 90).

11. Keeping firmly in mind our observation of ch. 24, that "ought is certainly not particularly a moral word." We need to beware of pre-anaytical meanings of words, such as this.

12. This may be so even where these sentences are imperative, interrogative, and exclamatory (rather than declarative).

13. This relates to our earlier discussion of *motivation*. When the world in my mind (Boolean 1) does not coincide with the world around me (Boolean 0), I am

Imagine a fictional example. I return home after work one evening to find that my house is not where it was in the morning. Instead, it stands ten yards to the right. In such a case, the feeling that "this ought not to be" or "I ought to see something else" is overwhelming. This reveals, too, that we thought all along that the house *ought* to be where it originally was.[14] In due course, we shall present some real examples.

In reality, we make value judgments at every turn—as to how things ought or ought not to be. I look for my keys on the table. "Where did I put my keys?" I ask myself. "They *ought* to be on the table." As I open the front door, it groans on its hinges. "I must buy some oil today," I say to myself. "These hinges *ought* not to sound that way." "Beautiful weather," I think, as I step out of the door—the sunshine on my face, and the fresh morning air. It *ought* to be this way—always. "Let me buy Julie flowers," I say to myself. "She *ought* to have a token of my love."[15]

All these things, and many more, express value judgments—namely, how relations between things, and the various arrangements of my life, ought or ought not to be.

Yet even this is too simple. We are immersed in a sea of relations, every moment of every day. Everywhere we look, we form innumerable judgments, as fast as our eyes absorb the view.

---

moved to act. Consider now the case where the world in my mind (Boolean 1) *does* coincide with the world around me (Boolean 1). In these terms, nothing changes in my mind as I shift from "is" to "ought."

14. Karl-Dieter Opp wrote, "There are different processes that lead to the emergence of norms. One process begins with a behavioral regularity. It is often claimed in the literature that if such a behavioral regularity has been established the respective behavior turns into a norm: what is becomes what ought to be" (Opp, "What Is," 13).

15. Again, someone might counter, "Ought is about things that are morally obligatory. It is not about keys and flowers and the like." Yet what, then, is morally obligatory? What is not? Where do we draw the line?

Chapter 26

Relating Science

*The exact sciences, although they seem to be exclusively
about fact, may be fully reconciled with value.*

WE HAVE SAID THAT the world as we experience it from day to day may all of it be interpreted in terms of *value*. Things are arranged as they ought to be or as they ought not to be. Even where they *are* arranged like this or like that, so they *ought* to be.[1]

So far, we have used ordinary, everyday examples to illustrate the point: moving a table in a room, buying Julie flowers, or stepping out into a sunny day. However, we have not yet applied such thinking to *science*—above all, that special case, the *exact* sciences: physics, chemistry, astronomy, and (arguably) mathematics.

Is it possible to interpret *science* in terms of value? The point is that, if we still must separate life into the exact sciences versus the rest of life, then we have not truly solved the fact-value problem. We are left with the language of hard science, on the one hand, against every other kind of language, on the other, and the status quo remains. We continue to live in a dichotomous world.

---

1. Someone might suggest an outlying example. "Your shirt is blue. That simply is." Yet depending on my usual style, there will always be some level of expectation. For example, neon pink or lemon yellow, in my own case, would surely elicit some comments.

To put it another way, there is a perception that the exact sciences lie beyond all value and are untouched by it. This attitude is typified by the chemist Antoine Lavoisier: "We must trust to nothing but facts. These are presented to us by Nature, and cannot deceive."[2] Philosophy professor Abdul Mannan described the attitude like this: "As far as natural science is concerned, there is no scope for any subjective elements. Scientific knowledge is purely objective, and it is an objective description of the real structure of the world."[3]

There are, of course, sciences that are not "exact" and sciences that are not wholly objective. Many academic disciplines have a large *subjective* component. That is to say, knowledge combines with human judgment and personality.

This is true of the social sciences, psychology, education, literature, the visual arts, and many other fields. All represent a *mix* of fact and value. Even those fields that are more definitely marked by quantitative expression are intuitively less than objective. Medicine, we feel, should be humane, finance people-friendly, architecture felicitous, and so on.

However, in the case of the exact sciences, we feel that these have little, if anything, to do with value. They are "exact" because they are capable of accurate quantitative expression, precise prediction, and rigorous methods of testing hypotheses. Their statements are true or false, right or wrong.

Philosophy professor Michael Luntley made the following distinction: the language of science allows only for the physical properties of things, while the language of mind has to do with perspective.[4] To put it simply, we have the language of science, on the one hand, and the language of mind, on the other. The language of science is objective, while the language of mind is subjective.

How, then, may we reconcile the language of science with the language of mind? We would do well to run over some basics, as to what science is.

---

2. Lavoisier, *Elements of Chemistry*, xviii. Lavoisier, apparently, was himself deceived. As an example, some have characterized his theory of acidity as a "major mistake" (Crosland, "Lavoisier's Theory of Acidity," para. 1).

3. Mannan, "Science and Subjectivity," para. 2. This is called *product* objectivity. Contrast with *process* objectivity, which states that the products of science are not tainted by human desires, goals, capabilities, or experience.

4. Luntley, *Reason, Truth and Self*, 56.

Relating Science   181

The "Ought" of Science

Consider those features of the exact sciences ("science," for short) that characterize it above all. Two things in particular separate science from all other academic disciplines—and for that matter, from all of our everyday thought.

Firstly, science typically proceeds by isolating individual mechanisms in the artificial circumstances of carefully controlled experiments[5]—called scientific control. All other academic disciplines—the human sciences, in particular—in one way or another admit a greater subjectivity.

That is, science must minimize unwanted influences on independent variables. In order for science to advance, we may need to exclude the smallest temperature variations, the tiniest impurities, the most elusive electrical interferences—even the smallest variations or uncertainties at the subatomic level.

Secondly, such tight control requires of us what we call a strictly normed language.[6] There is no room in science for terms that are vague or ambiguous. While in everyday conversation, we may use words fairly carelessly or loosely, science requires complete and explicit agreement of terms. Without this, it cannot progress.

We might think that these two features—scientific control, on the one hand, and a strictly normed language, on the other—make science an objective pursuit above all others. But this is a view that is less trusted today than it once was. The exact sciences are not what we once thought they were.[7]

Particularly in the wake of the great shift from Newtonian physics to Einsteinian physics and quantum field theory at the beginning of the twentieth century, the person in the street has become aware that the laws of nature may, so to speak, be *overturned*—or, to put it more conservatively, *modified*.[8] Their laws are no longer immutable.

---

5. "Scientists ideate adequate experiments to isolate new pieces of knowledge" (Redaelli, "Guide to Using," second to last para).

6. The philosopher and logician Norman L. Thomas wrote, "Intuitive meanings are suppressed and explicit meanings are formulated and clarified with as much precision as possible" (N. Thomas, *Modern Logic*, 59).

7. Laws and theories have much in common. Usually, it is said that theories are broader in scope than laws: they explain laws.

8. The philosopher Hans Vaihinger famously stated that most scientific concepts are adopted by scientists "as if" they are true (Vaihinger, *Philosophy of "As If,"* 91).

> Different hypotheses are sometimes offered for one and the same motion. . . . These hypotheses need not be true nor even probable.
>
> —ANDREAS OSIANDER[9]

We now know that Newtonian physics was an approximation of more accurate forms. The examples are many, but the best may lie in our Global Positioning System (GPS). Without the application of Einsteinian physics, this system would fail within minutes. There would be, for a while at least, global chaos.[10]

Shifts in scientific theory have changed our *perception* of theory. The philosophers of science Hanne Andersen and Brian Hepburn wrote that there have been "changes in beliefs about the certainty or fallibility of scientific knowledge."[11] Describing the situation today, Mel Thompson wrote, "All scientific theories are provisional and limited."[12]

What, then, do we mean by "scientific theories"? In this book, scientific theories have two aspects, which are closely related to one other:

- Firstly, there are the theories themselves and what they *include*.[13]

- Secondly, there are the *bounds* of such theories, which is their scope—and, most importantly, what lies *beyond* that scope.

To put it another way, we need to examine both what theory *is* and what it is not.

Various philosophers have held a similar view, among them Charles Sanders Peirce and William James.

9. Bertrand Russell wrote, "In more scientific matters, it is certain that there are often two or more hypotheses which account for all the known facts on the same subject" (Russell, *Problems of Philosophy*, 71). Richard Feynman wrote, "One of the amazing characteristics of nature is this variety of interpretational schemes" (Wolchover "Different Kind of Theory").

10. Will, "Einstein's Relativity," second to last para.

11. Hepburn and Andersen, "Scientific Method," §3.

12. Thompson, *Philosophy for Life*. Norman L. Thomas wrote that the postulates and axioms of mathematics and logic are "simply assumptions . . . that will generally prove to be useful" (N. Thomas, *Modern Logic*, 12).

13. This includes the *elements* of theory: concepts, variables, statements, and formats (Sundas, "Unit-2 Theory and Research," 24–25).

Relating Science 183

What Theory Is

The fact that scientific theories are now judged pragmatically would suggest that, rather than representing fact, they are what we think they ought to be, at any given time. No longer do we ask whether they are true. Rather, we ask whether they work.

We shall suggest, then, that science, when it works as we expect it to do, is as we think it ought to be. It traces relations as we think it *should*.[14] Conversely, science, when it is superseded, or when a hypothesis, even a theory, fails, it *is not* as we think it ought to be. It does *not* trace relations as we think it should.[15]

But wait a moment, someone might say. This may be true of *frontier* science, not of *normal* science. We may expect that, when we are at the *frontiers* of science, scientists will think that things ought or ought not to be a certain way—yet the laws of *normal* science are well established. There is no question that they are true. For instance:

- $a = \Delta v \div \Delta t$ (acceleration)
- $dL \div dT = \alpha L$ (thermal expansion)[16]
- $e = mc^2$ (relativity)

Here we find—or *suppose* that we find—no value judgments at all. How should subjectivity creep into straightforward equations?

We shall begin by focusing on the kind of science where the "ought" is most pronounced. It is indeed *frontier* science. We shall consider pivotal moments in the areas of nuclear physics and medicine, then consider whether there is anything there that applies to more ordinary science.

In the first example, the physicist Ernest Rutherford wrote in his notebook his astonishment at the results of his 1911 gold foil experiment, in

---

14. Various thinkers have described science in terms of tracing relations. For instance, Immanuel Kant suggested that those theories are most acceptable that extend our knowledge of relations among phenomena (Losee, *Historical Introduction*, 98). The physicist and philosopher Ernst Mach noted that the scientist seeks to formulate relations that summarize great numbers of facts (Losee, *Historical Introduction*, 146).

15. Of course, we assume that nature itself does not change, and this seems to be a reasonable assumption. It is, however, the *laws* that we apply to it that do.

16. This is often simplified to $\Delta L \approx \alpha L \Delta T$. Generally speaking, some equations are said to be mathematical, some scientific, and some between the two. To put it too simply, scientific equations purport to describe the real world, while mathematical equations relate to nothing but themselves.

which alpha particles unexpectedly bounced off the foil. He noted, "It was quite the most incredible event that has ever happened to me in my life."[17]

Had Rutherford merely been *toying* with the experiment, without reference to a theory to which he was deeply committed,[18] the surprise may not have been great. But his results defied belief. It *ought* not to have happened that way. He *expected* something else. And on recognizing that, indeed, it ought to be that way, he birthed the sciences of nuclear and atomic physics.

Another example is that of professors Ioannis Yannas and John Burke, the pioneers of synthetic skin. Their greatest breakthrough, in 1981, was "a big surprise, totally unexpected, and hard to believe."[19] After testing various artificial membranes to speed up the healing of the skin, they found one that, contrariwise, slowed it down. Appearing at first to fail in its purpose, it regenerated the skin. This changed the treatment of burns and saved innumerable lives. Again, it *ought* not to have happened that way—until Yannas and Burke realized that, indeed, it ought.[20]

In fact, scientists continually speak in terms of "ought" —something that we briefly surveyed in chapter 24. This was all the more so in the decades before the Newtonian paradigm was, in a sense, superseded.[21] More and more, scientists were seeing things that both ought and ought not to be.

Now if we broaden our view to survey the science of the past and imagine the science of the future, we realize that the way we think that science *ought* to be is subject to change. The way that we now arrange the physical world in our minds is not the way that we arranged it in the past,

---

17. Rutherford, "Rutherford's Incredible Event," 1.

18. Rutherford had "formed preconceptions," writes Chem Talk ("Discovering the Nucleus," para. 5).

19. See MIT Mechanical Engineering, "Life-Saving Discovery," 0:46.

20. We find an opposite example in the polymath Galileo Galilei, who puzzlingly recorded incorrect data for the period of a pendulum. Yet his data fitted with that which he *ought* to have found at the time. Whatever he was thinking, Galileo recorded the data which ought to be true but was not—and the data which he thought to be true was not what it ought to be (Morgan, "Galileo's Pendulum Experiments," §5).

21. Professor of physics David Lyth put it like this: "Einstein's general theory of relativity killed off common-sense physics" (Lyth, "How Einstein's General Theory").

nor will it be the way that we arrange it in the future.[22] Science, at any given time, is merely as it ought to be.[23]

What, then, of *normal* science—established science—familiar science, which we use from day to day? How should *that* have to do with ought and expectation?

When we reformulate the matter, there is something in frontier science that applies to more ordinary science, too. Not only do our theories shift over time, but they never apply to the world in the first place, in quite the perfect way.

What Theory Is Not

Even if science appears to be a perfect match with reality, in terms of the most stringent requirements of the scientific method—that is, if its theories appear to be completely accurate as confirmed in carefully controlled experiments—they do not represent a complete or perfect fit with reality.

Suppose that we take Ohm's law. On the surface of it, this is verified every time we turn on a light, use a computer, or do a myriad of things—all of which work, not only inside but outside the laboratory.

Yet Ohm's law nowhere applies perfectly to any given system or device. This is not because of Ohm's law but because of the system or device. If one calculates R (Ohms) on the basis of a resistor's nominal value,[24] one may come only within 2 percent of the correct answer, in cases where one uses *precision* components—and that is without reckoning with environmental influences or component failures.

We may formulate the issue like this: we develop theories within closed systems, while reality is an *open* system. We call it a "global system," which is really no system at all. Our theory never quite fits the reality.[25]

The problem, however, is bigger than this. The fact that theory (a closed system) is often enough introduced into the *world* (a global

---

22. In fact, science always represents an imperfect fit with reality. This cannot in the strictest sense be characterized as "is."

23. Albert Einstein wrote that "every theory is speculative," often "precarious" (Einstein, *Ideas and Opinions*, 349).

24. Usually, one depends on nominal values—yet even if one measured R (almost) precisely, its value would change, due to various influences.

25. Science editor Richard Webb writes, "Why do the abstractions of mathematics match reality? Indeed, do they really match, or is it all an illusion?" (Webb, *How Numbers Work*, 183).

system) does not merely give rise to theoretical niggles. The mismatch means that the theory left something out—and that something is often of enormous importance.

Let us take a real-world example, which relates to thermal expansion. As temperatures rose in the United Kingdom towards the turn of the century, railway tracks all over the nation began to *buckle*. In some cases, trains were rerouted or rescheduled, while, in others, they greatly reduced their speed. Commuters stood on platforms in the heat, wondering what had happened to the timetables.[26]

The answer was, in an important sense, a philosophical one. The laws of thermal expansion had been applied to railway tracks without taking environmental conditions into account—at least, not those that counted at that time. The tracks had been affected by climate change.

The point is that, in this way or that, the world will often encompass things that our theories do not. There are things in the world that we do not find in the theories we *apply* to the world. This is because a scientific theory encompasses just a few relations among things, amidst an endless expanse of relations.[27]

Interestingly, this comes close to the definition of "subjective": "belonging to, proceeding from, or relating to the mind of the thinking subject and not the nature of the object being considered."[28]

It is the heresy of Plato's forms. Ordinary things, held Plato, imitate *forms*—abstract objects, which exist independently of thought.[29] We hold such forms up to our reality—in this case, theories, equations, algorithms, and so on—and, more often than not, we assume that this is how the world *is*. In fact, we often speak of our forms as "laws" of nature. Certainly, we often qualify this—with the amendment "under the given conditions" (in Latin, *ceteris paribus*)—yet we generally take our laws *as read*.[30]

---

26. Dobney, "Future Cost," 25.

27. The theoretical physicist Pierre Duhem noted, "Agreement with experiment is the sole criterion of truth for a physical theory" (Duhem, *Aim and Structure*, 21). Contrast this with my statement: "There are things in the world that we do not find in the theories we *apply* to the world."

28. "Subjective," in *Collins English Dictionary*.

29. There are, however, various definitions and interpretations of Plato's forms, some of which are quite complex.

30. We supposedly overcame Plato's heresy through the astronomer Johannes Kepler's insight, "Mathematics wasn't at the root of the physical laws governing nature" (Siegel, "No, the Universe").

They are not laws—at least, not in the world in which we live. They are only laws in the fleeting, artificial conditions of the laboratory. The philosopher of science Stephen Toulmin described the laws of science as "idealized, imaginary states-of-affairs that never in practice occur."[31] If they "never in practice occur," then they are the imaginings of the thinking subject—at least, to some extent, which may actually matter.

Our science so often seems like a bottle top which does not quite fit the bottle. One can try to *force* the bottle top into place, but one might damage the top—or damage the bottle—or the bottle may no longer serve its purpose of sealing the liquid inside.[32] Somewhere along the line, there may be consequences—and there often are.

Excursus

Someone might counter, "Do we really have to take things as seriously as this? We made a *mistake* about the rails. We misjudged the matter. We didn't see it coming. It's as simple as that. We can fix it."

Yet it lies in the very nature of modern knowledge to exclude the whole. We do this through minimizing unwanted influences on independent variables, which is the scientific method. Buckled tracks are merely one among many symptoms of the same.

This means that we should think much further than merely fixing the problems in front of us. Often, the fix does not lie within the system itself. In the case of buckled tracks, it lies both in the system and in all the world.

This leads us to yet another problem, which is closely related, if not identical to the above. It is, in the words of Simon Blackburn, "the selection of particular facts as the essential ones."[33] The essential ones are those that fall under our pure and perfect equations.

---

31. Toulmin, quoted in Cleveland, *History of Energy*, 91. The physicist Percy W. Bridgman wrote, "No empirical science can ever make exact statements" (Bridgman, *Modern Physics*, 34).

32. The physicist and astronomer Arthur Eddington wrote, "The mind has by its selective power fitted the processes of Nature into a frame of law of a pattern largely of its own choosing" (Eddington, *Nature of Physical World*, 53).

33. Blackburn, "Fact/Value Distinction," in *ODP*, 134.

Wherever we think "system"—in our last example, a system of tracks and trains—we must first *scope* the system. We must first define its boundaries. We must choose what it will include and what not.[34] Yet this is practically impossible—because, as Thomas Berry, a historian of the Earth, put it, "nothing is completely itself without everything else."[35]

Consider what this means in the context of a philosophy of relations. We may augment a theoretical system as we take additional factors on board, but if we do not stop at some point, we overrun the boundaries of a practical system.

In the case of railway tracks, we may (credibly) relate them to carbon emissions, trackside vegetation, even British politics—in fact, to everything in the world, when we come to think of it: the production of cement in South Africa, deforestation in the Congo, cattle in the Alps, and a million things besides.

Yet we must *decide*—and, in many cases, we merely *assume* that we know—what really matters to our system. For British Rail, global warming did not matter—not, at least, over a given period of time.

When one takes a wife or a husband; when one does a business deal, when one buys an item at a store; and, indeed, when one does science, one often discovers that the *ideal* thing was not the *real* thing. One calls it subjectivity, partiality, blindness. For now, therefore, we claim that science may be catalogued under value. However, this is not all that there is to say about the subectivity of science. Our most important observations are held over for chapter 36, on the purpose of relations.

---

34. Such systems may be physical but also conceptual or hybrid (which is both). They may also be human-made or natural. Even if natural, they are, in a sense, human-made, since they are described by humans.

35. Berry, *Dream of the Earth*, 91. This would apply to all knowledge, not merely scientific knowledge.

Chapter 27

Relating Math

Mathematics, too, may be understood entirely in terms of value. Though perfect in isolation, math is all about value as soon as it is applied.

As with science, so with mathematics. If we still must separate life into mathematics versus the rest of language, then we have failed to truly bridge the fact-value divide. That is, if we trace relations between different things on different levels without a means to unite them—in this case, mathematical things with other things—our reality is still split in two or more.

Mathematics has a close relationship with the exact sciences and may or may not be viewed as one of them. If science is important to the consideration of the fact-value distinction, then so, too, is mathematics. We must relate mathematics to value—if, indeed, this can be done.

This is important not merely in the quest to unite all knowledge and experience in a single metaphysics. It is important because, if we should find that mathematics can and should be brought under the concept of "ought," then mathematics must be *subjective*—and if mathematics is subjective, then we would make a big mistake in treating it otherwise.

We may compare it with the soldier who believes he carries an accurate weapon, where, in reality, it is less than true and may soon be the end of him. If mathematics is tinged—or, worse, suffused—with value, it becomes a hazardous tool.

Typically, mathematics, even more so than science, is thought to be supremely objective. Here, there are no failed experiments, no false interpretations, no paradigm shifts, no danger that our sums may be overthrown.[1] On the surface of it, mathematical statements would seem to present us with the very opposite of value judgments:

1 + 1 = 2
the logarithm of 1 = 0
the square root of 1 = 1

and so on.

But here lies the problem. By and large, we *apply* mathematics to things in the real world[2]—and, in the real world, we first need to identify units of quantity. This begins with *quantification*, which maps our human sense observations into units of quantity.

Now it may seem to us, on first impression, that units of quantity come ready made: apples come in ones, and so do oranges, people, animals, days, nights, doors, windows, and a great deal more. On this basis, we count things and perform various mathematical operations on them.

On closer examination, however, mathematics is not that clean cut. As soon as we give *content* to the units of math—say, gallons in a tank, ships on the sea, even clouds with noses—that which is *contained* in the numbers takes on a vital importance.

Albert Einstein considered that a unit "singles out a complex from nature."[3] That is, one takes a complex of things and defines it as one. This means that various things may hide inside one and the same unit.[4] Mathematics, wrote the mathematician Luitzen E. J. Brouwer, consists in *constructing* the objects with which it works.[5]

1. Godfrey Vesey and Paul Foulkes summarize it like this: "Two twos make four and no mistake" (Vesey and Foulkes, "Mathematics," in *COLDP*, 183–84)—with the exception of mathematical induction.

2. The exception, of course, being pure mathematics, which Georg Cantor called "free mathematics" ("Synopsis of Georg Cantor"). Mathematics, without all application, would be irrelevant.

3. Einstein, quoted in Douglas, *Quotes of Albert Einstein*, 60. The mathematician Georg Cantor wrote that individual variables contain "definite, distinguishable objects of perception or thought conceived as a whole" (Faulkner and Hosch, *Numbers and Measurements*, 74).

4. Thus the possibilities for counting and calculating things become boundless, because there is no end to the possible complexes we can create and designate as units. In fact, a unit can contain anything.

5. Vesey and Foulkes, "Mathematics," in *COLDP*, 183–84.

Relating Math 191

But now, if units are complexes, as Einstein suggests, what *ought* they to include and what not?[6] Further, how *true* are they? How suited are they to manipulation with mathematical formulae or models?[7] Thomas Mautner, the editor of the *Penguin Dictionary of Philosophy*, cautioned, "All abstraction involves some falsification."[8]

Suppose my kitchen hand is making a sauce for some dinner guests. He knows that he derives the perfect taste with, say, two-thirds button mushrooms and one-third cherry tomatoes. With ten button mushrooms in hand, 10 ÷ ⅔ - 10 = 5 cherry tomatoes—or however one might wish to calculate it.

But now he makes a blunder, which has dramatic effects. He forgets that one of the guests is allergic to button mushrooms. While the mathematics works perfectly for the perfect sauce, it strips out *context*. If it is not the guest with an allergy, then it is something else. These mushrooms may have been grown, say, in radioactive soil or stolen from a poor man's greenhouse. Therefore, our calculations frequently *miss* something in the real world.

Of course, it is precisely because mathematics strips out context that it is so powerful. At the same time, this makes it *dangerously* powerful. Even if we hold that it cannot be tainted in the isolation of its own perfection, the harms that *result* from it cannot happen without it.

Parallel Worlds

The illustration of mushrooms and tomatoes is limited in its usefulness, yet it does make this point: while the formula may be true in itself, it may leave things out. And where, in this case, the application of a formula may make someone ill in a restaurant, in many cases, it may bedevil the whole planet. Take, say, the various formulae that were used in the development of the internal combustion engine. They left out the planet.[9]

---

6. We not only scope the system, but we scope the unit.

7. Physics professor Chad Orzel notes, "Empirically, every physical phenomenon we've been able to test is governed by simple mathematical rules, but it's not obvious that that needs to be true" (Orzel, "What Are the Limits").

8. Mautner, "Abstract," in *PDP*, 3. Percy W. Bridgman considered that the problem lay, in a sense, in mathematics itself: "If the statement of arithmetic is to be an exact statement in the mathematical sense . . . this sort of thing is never experienced" (Bridgman, *Modern Physics*, 34).

9. This touches on chemistry in particular. The most basic reaction of the internal

The mathematical models we first applied to nuclear fission did not factor in radioactive fallout,[10] nor did the models we first applied to economics factor in human behavior, nor did the models we first applied to pesticides factor in nontarget zones. We omitted crucial sums. The examples are legion and, in most cases, more subtle than this.

Another way of putting it is that we are accustomed to applying mathematics—as we do with science—to closed systems. Within closed systems, the mathematics usually works—or works well *enough*. However, we live in a *global* system, which is the world. The statistician George Box reportedly said: "All models are wrong, but some are useful."[11]

Is nature's reality deeper than mathematics itself?

—KITTY FERGUSON

Many have imagined that mathematics represents a perfect fit with our world. Galileo suggested this when he wrote, "Mathematics is the language in which God has written the universe."[12] Vesey and Foulkes summed it up, "The world is itself constructed on rational lines."[13] This suggests that our mathematical equations and mathematical models (which are *sets* of equations) reveal the secrets of the entire cosmos.

The reality, however, is quite different. Our world today is carved up into several major mathematical models: classical mechanics, relativistic mechanics, quantum mechanics, and quantum field theory. Further, we find a profusion of smaller models, including dynamical systems, statistical models, differential equations, and game theoretic models. These

---

combustion engine is $C_8H_{18} + 12.5\ O_2 \rightarrow 8\ CO_2 + 9\ H_2O$.

10. Perhaps, seriously underestimated it (Taylor, "Nevada Test Site Downwinders," para. 10). Cellular biologist Barry Commoner wrote, "The greatest single cause of environmental contamination of this planet is radioactivity from test explosions of nuclear weapons in the atmosphere" (Commoner, "Fallout and Water Pollution," 2).

11. For a similar statement, see Box, "Science and Statistics," 792. Similarly, the work of cognitive scientist and philosopher Aaron Sloman has been summarized as "Theories distort in order to simplify" (See Sloman, *Computer Revolution in Philosophy*, 6.)

12. Croy, *Galileo Galilei*, 8.

13. Vesey and Foulkes, "Rationalism," in *COLDP*, 248. Euclid considered that laws of procedure were "woven into the cloth of nature" (N. Thomas, *Modern Logic*, 8).

may be applied, in turn, in diverse fields: science, technology, business, engineering, sports, and so on.

Mathematics, therefore, is applied in those areas where it *works* or *seems* to work—which is to say, where the relations it traces seem to be true. However, it is neither a unified nor an ideal system.[14]

An Incompleteness Theorem

It is often the case that, over time, we discover a lack of agreement between the mathematics and our experimental measurements.

The theory of circular orbits and epicycles did not fully explain the motion of the planets and was supplanted by a model of elliptical orbits. Scalars proved to be inadequate for fully understanding forces and were superseded by vectors, then tensors. Classical thermodynamics did not fully explain flow rates and was superseded by generalized thermodynamics.

In all these examples and more, the mathematics merely expressed what scientists considered was the case at the time. Scientists, it turned out, were not talking about what *was* the case. In terms of this metaphysics, they were talking about what *ought* to be the case, where, again and again, what ought to be the case was discarded or superseded.[15]

The theoretical physicist Richard Feynman observed that mathematicians deal only with the *structure* of reasoning and do not really care what they are talking *about*.[16] In fact, our reality may not be mathematical. Albert Einstein wrote, "As far as the laws of mathematics refer to reality, they are not certain; and as far as they are certain, they do not refer to reality."[17]

The question now arises, as it will with regard to science in chapter 30: where there is a lack of agreement between the math and experimental

---

14. Computer science professor Pedro Domingos wrote, in the context of theory, "Science today is thoroughly balkanized." (Domingos, *Master Algorithm*, 46).

15. Note that we do not say that "ought" is integral to mathematics. One does not say, for example, "One plus one *ought* to be two." At least, this is debatable.

16. Feynman, *Character of Physical Law*, 55. Alfred North Whitehead wrote, "There are no whole truths; all truths are half-truths. It is trying to treat them as whole truths that plays the devil" (Ratcliffe, *Concise Oxford Dictionary*, 11.66).

17. Berlinghoff, *Mathematics Sampler*, 282. Philosopher of science Mary Hesse wrote, "No field of inquiry is ever closed in such a way that its formal description exhausts all that physics ever wants to say about it" (M. Hesse, *Forces and Fields*, 25).

measurements, is it the math that is responsible—with all the trouble that ensues—or does it come down to the *application* of math?

It is clear that some things in this world are unmeasurable, and some things are incalculable.[18] Some, too, are unmeasurable in some *degree*, and some are calculable with different levels of confidence. In some such cases, when we use math—or use certain kinds of math—we make a mistake.

Yet we do know, further, that some of the trouble lies not so much with us as with math itself.[19] The mathematician Rodney D. Holder noted that, with regard to numbers, "a finite number of decimal places constitutes an error,"[20] while Kurt Gödel pointed to the incompleteness of number *systems*—given that what this really means is "ill-understood."[21]

Mathematics, in fact, may give us special insight into the is-ought distinction. Consider that negative results often represent *deficiencies*. We do our sums, only to discover that we have a deficit, a shortfall, or arrears. We calculate that our fuel will run out too soon, our signal is too weak, or we have underbudgeted for a mission. Or, on the other hand, positive results often indicate excess, wastage, or residue. We paid too much, or we overloaded the plane, and so on. Things *ought* not to be that way.

Unless, that is, we obtain a nil result. Nil often reflects that all expectations are met. A buyer is satisfied, and so is a seller; a mission is on target; projections are met. The Egyptian symbol for nil was *nfr*, meaning beautiful.[22] It is a beautiful moment, when both negative and positive oughts are united in one.

There is another connection, too, with fact and value. In chapter 25, we said that the fact-value distinction rests on David Hume's view of

---

18. Norman L. Thomas wrote that, when we apply deductive systems, "we know we cannot get exact measurements and relationships" (N. Thomas, *Modern Logic*, 152).

19. Percy W. Bridgman wrote, "It is the mathematics made by us which is imperfect and not our knowledge of nature" (Bridgman, *Modern Physics*, 62).

20. This error "propagates so rapidly that prediction is impossible" (Holder, *Nothing but Atoms*, 114). Owners of early Sinclair calculators, such as myself, viewed the propagation of errors with astonishment. While calculators are now much refined, the problem is still there and always will be. "Even the best device will never measure anything physical with infinite accuracy" (Webb, *How Numbers Work*, 164).

21. Wolchover, "How Gödel's Proof Works," para. 5. One of the better known consequences of Gödel's theorems is that "no program can find all the viruses on your computer" (Dawson, "Gödel and the Limits"). Consider, as a matter of interest, that no system in itself can prove one's veracity.

22. Bidny, "History of Zero," para. 2.

relations of objects. Things may be arranged as we think they ought to be or not. Now we see that this same view is embedded in mathematics, as follows:

Excursus

The biggest problem, however, lies unseen. Which is to say, it lies less in the math than it does in what is *missing* from the math.[23] Every system of mathematics has a ruthless disregard for the whole—a whole that lies outside one's sums.

We may express this mathematically through the following analogy:

- mathematics uses (chiefly) variables and operators
- any variable represents the whole world minus *not-the-variable*
- now assume that W = the whole (which is all possible variables)
- assume further that ø = any given operator (+ - × ÷ and so on)
- x + y = z now reads (W - not-x) ø (W - not-y) = W - not-z
- now remove the whole, W, from the equation

x, y, and z now become the results of unstated subtractions from an unknown variable. One has lost one's reference, W.[24] Yet if there is the slightest error in what one estimates W to be, one's math is quite likely compromised.[25]

---

23. Percy W. Bridgman pointed out, "There is another aspect of the use of mathematics in describing nature that is often lost sight of; namely, that any system of equations can contain only a very small part of the actual physical situation" (Bridgman, *Modern Physics*, 64).

24. In fact, x + y = z revolts against the whole, because it assumes a unitary result z, which treats itself as the whole. *As humans*, we can see that many things lie outside of x ø y = z, but if the equation could speak, it would report that it knows nothing of it.

25. Thus we omit a *pre-calculation*—and, in so doing, have transgressed our sixth ethical maxim, if not others: "Be guided by a deep sense of inadequacy." We return to Ohm. We have, as an example, two identical resistors in series, R1 × 2. The mathematical result can only be correct—namely, 2R—yet this fails to take into account either the internal characteristics of the resistors, or the external influences upon them, which lie in W.

Mathematics speaks of variables, on the one hand (which is things), and operators, on the other (which is relations). Therefore, it is a kind of *mathematical* thinking that lies at the core of the fact-value distinction and, with that, at the root of all our ethical disorientation today. Firstly, our sums falsely slip the assumption of "is" into our arrangement of things, and secondly (we expand on this in later chapters), they assume that things are complete in themselves.

Chapter 28

Relating Languages

There are at least three kinds of language: natural, scientific, and mathematical. When we understand individuation, we unify these languages.

WE HAVE SURVEYED THREE kinds of language: natural language, scientific language, and mathematical language. Our ordinary, familiar language, in the words of the linguist Lincoln Barnett, is "free-ranging,"[1] while the language of science is strictly normed, and the language of mathematics is—so it would seem—altogether abstract.

Yet as we survey these three "languages," we find something both curious and interesting about them. We freely mix them all in our daily conversation.

"I have two kinds of rope," says the owner of the hardware store. "Nylon and manila. I would go with nylon," he says. "See this chart here. Nylon has a high tensile strength, twice that of manila. You'll take nylon? That'll be eight dollars. Give me that tenner and I'll give you two. Have a good day." One has just used all three kinds of language rolled into one.

So far, we have sought to show that all three languages—ordinary language, scientific language, and mathematics—fall on the same side of the great fact-value divide. All three, we have said, describe how things

---

1. Barnett, *Treasure of Our Tongue*, 213.

ought or ought not to be related. Yet even when we have reconciled them in this way, they refer to different *kinds* of things.[2]

The rope that I feel in my hands seems quite different to the tensile strength that flickers on a laboratory monitor, and that again seems quite different from the ten minus two in the shopkeeper's head.

How do we reconcile these different *kinds* of things? How shall we unite them as one?

In fact, the problem is not difficult to solve, when we resort once more to a low-level concept—in this case, the concept of *individuation*. To individuate is both to define things and to separate them from other things.

Individuation and Relations

Our reality has been variously described as an undifferentiated stream of experience, a kaleidoscopic flux of impressions, or a swirling cloud without determinate shape. To make sense of it, then, we need to separate this undifferentiated stream—sounds and sights, surfaces and motions—into individual units.

From our earliest childhood, we begin to individuate people, playthings, sounds, smells, and many things besides. We single them out. Before long, we begin to look at *picture* books, in which individuated things are represented in pictures—their names typically being printed underneath them: apple, orange, dog, cat, sun, moon, and so on.

Now we may note that none of these things can exist in *isolation*. An apple must at some time be attached to a tree, a dog must breathe air, a moon needs a planet, and so on.

Yet during the process of individuation, we strip away these interconnections or interdependencies, and seek to create something that is self-contained. In Hegelian-style philosophy, we say that such individuated things are *abstract*—because we think of them in isolation from the whole to which they belong.[3]

---

2. There may, of course, be more than three kinds of things. Ferdinand de Saussure, as an example, wrote about "special languages" other than those covered here (Saussure, *Course in General Linguistics*, 21). This is readily reconciled with this metaphysics.

3. Mautner, "Abstract," in *PDP*, 3.

> An immense number of associations are formed, and remain as long as they endure, in the background of consciousness.
> —CHARLES SANDERS PEIRCE

Consider a dog. When we speak of a dog as an individuated thing, we have little interest in what it eats, if it sleeps, or even whether it has four legs or three. It is something else that makes it a "dog." Or consider an apple. We have little interest in the tree to which it once was attached, the climate that nurtured it, or the soil in which it grew. It is something else that makes it an "apple." A dog is a dog in itself; an apple is an apple in itself.

To put it another way, when we individuate something, it loses much of its *informational content*. We make great subtractions from the total concept. In reality, it is impossible to imagine a dog without air, food, or something to stand on—or an apple without a tree and all that nurtures it and makes it grow. The individuated "dog" and the individuated "apple" need none of this.[4]

Yet even at the same time, we carry all the *associations* of individuated things in the back of our minds—or, rather, all the associations we have *made*, which we retain in our minds. They are there, even as we exclude them.

Put it this way. We allow some aspects of a thing to recede yet not to leave the picture—much as happens when we pass a magnifying glass over a newspaper. While some of the page jumps out at us through the glass, the rest is still there.[5] Both the focus and the fringe remain (mostly) within view.

This is easy to prove. Consider the sentence, "He threw the apple into the furnace." We all know that we shall never eat this apple, yet the individuated "apple" does not contain such information. Or consider the sentence, "The dog yelped." We all know that something happened to hurt the dog, yet the individuated "dog" does not reveal it.[6]

---

4. There is a twist on the theme. Suppose that the dog or the apple is red. We strip away everything but this redness. What do we have left? Objects? Properties? In terms of ch. 3, we have *things*, no less than the things "dog" or "apple."

5. The polymath Michael Polanyi established that we are typically aware of certain objects without our attention being fixed on them (Polanyi, *Study of Man*, 29–30).

6. Pedro Domingos noted that, in many cases, "the best way to understand an

When we understand both ordinary things and scientific things in terms of *abstraction*, we find that we are in a position to answer the question as to what unites our natural language with our scientific language.[7]

Our ordinary, everyday language individuates things by stripping away relations. So, too, does our scientific language, which excludes unwanted influences on independent variables. In an important sense, therefore, we "do science" all day long. Every time we speak a word—whether it be an ordinary word or a scientific term—there is little difference in the most fundamental, general way.[8]

Paul Hoyningen-Huene, a neo-Kantian philosopher, questioned the nature of science as a special activity. "Scientific knowledge," he wrote, "differs from other kinds of knowledge, especially everyday knowledge, primarily by being more systematic."[9] In other words, the difference is not a matter of kind but of degree. While one cannot *equate* the processes of abstraction in natural language and science, they are closely related to one another.

In both cases, we *individuate* things—except that, in the case of science, we are more rigorous in putting out of our minds the relations we do not need—or *think* we do not need.

Mathematical Abstraction

What, then, of mathematics?

Here we find another kind of abstraction, which now strips away (almost) everything from the relations in which a thing is involved—in fact, strips away the very *names* of things. Aristotle wrote that, through

---

entity . . . is to understand how it relates to other entities" (Domingos, *Master Algorithm*, 228).

7. Notice that this is not merely about words, It shows us how we think about reality. The things that we call "concrete" are really not that concrete at all. In the same way that words may be distributed or dissipated through space and time, so may things.

8. Godfrey Vesey and Paul Foulkes observed, "One may as well ask whether in the end there are many logics, all independent from each other, or whether they all somehow depend on the logic of ordinary discourse. Here we have yet another controversial issue that cannot be resolved in terms of logical operations within the systems concerned" (Vesey and Foulkes, "Logic," in *COLDP*, 174–76).

9. Hoyningen-Huene, *Systematicity*, 14.

the process of mathematical abstraction, one "strips off all the sensible qualities... and leaves only the quantitative and continuous."[10]

A dog may become a mere number. We strip away everything except the fact that it is "one" thing. An apple may be reduced to its weight alone. We strip away everything except the fact that it weighs, say, one hundred grams. We may do the same with *events*. We turn them into *units*: "They held three concerts this week," or "We are pumping ten gallons per minute."

Anthony Grayling, a philosopher and ethicist, summed it up: "Individuation provides not only a criterion for discriminating among particulars, but for counting them."[11] Therefore, the process by which the mathematician arrives at abstractions is very similar to the process for all abstractions.

On the surface of it, then, when we do mathematics, all that remains is the bare existence of an object or event, nothing more. The numeral 1 now indicates the existence of a single something that has no properties left. Thus we may *number* things or quantify them. I picked seven apples this week, but threw one into the furnace. $7 - 1 = 6$. I may further call the apples x, and say that $x = 7$ and $y = 1$, where y is all apples that went into the furnace, and so on.

By the time we have stripped off all sensible qualities, so reducing things to the level of science and, finally, mathematics, we might imagine that we are left with total abstraction. Yet even now, some content remains. It seems something like the background radiation of the universe—the telltale leftovers of something that once was.

It begins with the very numbers that we use. For example, we are more likely to use natural numbers for counting things and rational numbers for constructing things. We therefore find that, although we have stripped things down to numbers, we are not left with something of a completely different order.

---

10. Aristotle, *Metaphysics* 11.3, quoted in Bedürftig and Murawski, *Philosophy of Mathematics*, 35.

11. Grayling, "Individuation."

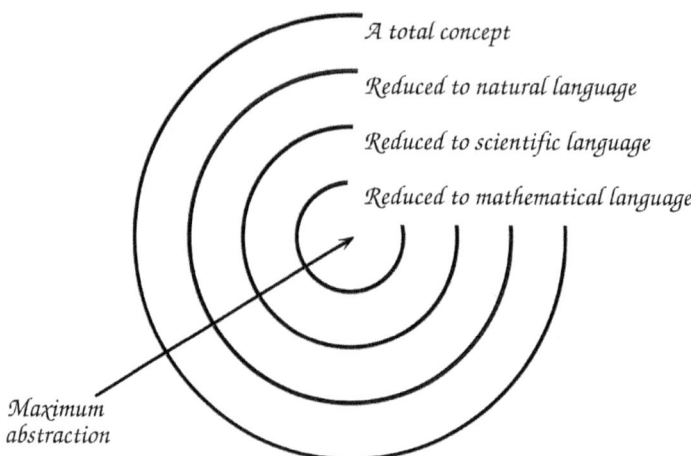

In addition to this, the way that we choose to *manipulate* numbers reflects the reality to which they are applied. Albert Einstein noted, "In practice the world of phenomena uniquely determines the theoretical system"—which is to say, the theoretical system is not empty, as it were.[12]

There is a *danger* in thinking of mathematics as a language set apart by itself—rather than a stripping-things-down, an abstraction—where mathematics really represents things that have been severed (if they really could be) from a great many relations in which they are involved. Whenever we assume that math is a stand-alone language, we may become overconfident of our sums. This may well end in hubris and disaster—to which we return shortly.

What now differentiates natural, scientific, and mathematical language is the *degree*—we might say *intensity*—of abstraction applied. All of our speech and all of our writing, will be found somewhere on a *continuum* of abstraction—or, as the diagram suggests, nearer or farther from the *core* of maximum abstraction.

---

12. Einstein, *Essays in Science*, 4. Bertrand Russell observed, "Although our general proposition '2 + 2 = 4' is *a priori*, all its applications . . . contain an empirical element" (Russell, *Problems of Philosophy*, 61). I propose that the general proposition itself contains an empirical element, not merely its applications.

Chapter 29

The Isolation of Science

Science advances by progressively screening things out. Yet in doing so, it dangerously isolates itself from a world of wider relations.

SCIENCE FILLS US WITH wonder. It has revolutionized almost every aspect of our lives, healing our bodies, improving our communications, introducing a multitude of comforts to our lives, and taking us places we once could only dream of—among many other things.

Yet at the same time, we have all become increasingly aware of the *dangers* of science. Some would say, the dangers of its *use*—a distinction upon which we shall reflect shortly.

A low-level idea usefully reveals both the limitations and the dangers of science. It is, again, the law of noncontradiction: A is *not* not-A. In keeping with this law, every variable that science puts into service *excludes* all others. The simplest of sums—take $x + y = z$—must cast away everything not-x, not-y, and not-z, not to speak of a and b and c, and every other variable under the sun.

To put it another way, if we accept that everything is related to everything else, then every time we do our sums, we *cut away* a part of our reality, if not most of it. Computer science professor Pedro Domingos wrote, "Even our most elaborate models are usually oversimplifications of reality."[1]

---

1. Domingos, *Master Algorithm*, 78. This oversimplification represents a great danger, causing many things to slip beyond humanity's control.

Now in terms of a philosophy of relations, we find a strange paradox. On the one hand, the relations between things are boundless; on the other hand, we must *strip away* relations if we are to do science. Science must screen out unwanted influences on independent variables.

Science—in particular, the exact sciences—requires the very opposite of openness to the totality of things, to survive and to thrive. It must reduce the number of things with which we work. This is its *sine qua non* (its "without-which-not"). It is the without-which-not, too, of all the phases of the scientific method: characterizations, hypotheses, predictions, and experimentation. The philosopher Mary Midgley called it the "austere modesty" of science.[2]

Now there is something that we may too easily miss. Not only does science screen things out; it also *screens things in*, so to speak. Not only must the scientific method screen out unwanted influences on independent variables; it should not exert any influence on influences it seeks to exclude. If it does, then scientific control is compromised.

For instance, we might seek to determine how much heat is required to convert a kilogram of ice into steam. But if, in the process, we should warm the laboratory as well, then the procedure is fundamentally flawed.

We may call this a "double isolation." Not only does science need to screen things out, but it needs to screen *itself* out from its environment—and this does not necessarily end at the conclusion of the experimental stage.

Physics investigates processes . . . by
progressively screening things out.
—WILHELM KAMLAH AND PAUL LORENZEN

Scientific Isolation

Science, after it has screened itself out from the world, ultimately needs to reenter the world. After the final, experimental stage of the scientific method, the artificial conditions of the laboratory are removed, and

---

2. Midgley, "Can Science Save," 24.

science begins again to have an effect on the world.³ This it does through engineering, then technology, which is the *application* of the discoveries of science.⁴

However, little thought is given to what occurs at this point.⁵ It seems much like catching some curious insects in a bottle. After studying their behavior in the bottle and exhausting our investigations of what they do there, we now release them into a new environment. Of course, we do not know quite what will happen next. They might die and return to the ground, or we might visit the same environment in fifty years' time, to discover that it has completely changed.

Similarly, science, when it reenters the world, may have effects on the world that go beyond anything that was formally taken into account in the scientific method. In the long run, these effects may impact an entire nation or span the globe.⁶

Science, therefore, must take account not only of the mechanisms that are isolated in controlled experiments but of those mechanisms that have been excluded from such control. These may potentially be infinite—yet there must be an attempt to account for them and list them. No experiment is truly complete, and no experimenter fully mature, until the experimenters have thought far beyond the laboratory.⁷

Today, some of our scientific pursuits are systematically regulated and supervised long after the final experimental stage of the scientific method—notably, in the areas of food, drugs, and (immediate) environmental impact, while a great many things remain ill considered, if they are considered at all. To put it simply, the scientific method needs

---

3. The philosopher Henri Bergson wrote, "The so-called isolated system remains subject to certain external influences. Science merely leaves these alone" (Bergson, *Creative Evolution*, 10).

4. Engineering and technology have been informally described as a question and answer: "How shall we accomplish this?" "*This* is how we accomplished it."

5. Much attention is given to certain *aspects* of what happens at this point, which usually has to do with profit.

6. Not only this, but technology may combine or accumulate the results of a wide range of experiments—in the construction of an airliner, for instance, a nuclear power station, or an irrigation network. Such combinations may have effects that are all but impossible to anticipate while studying the parts.

7. The sociologist Daniel Bell held that a post-industrial society would develop long-term forecasting of the effects of technological innovation (Bell, *Coming of Post-Industrial Society*, 199).

to explode its traditional bounds, to become an open-ended process that includes its widest possible implications.[8]

> ### Excursus
>
> It might seem odd to propose that every theory must be open to *invalidation*—that is, open to being found unsound or unacceptable in a global system.
>
> The fact is that no theory can be verified in a wider context. It must always, at some point, fail. The questions are how and how much. Here the danger lies, when theories are applied to the real world.
>
> How should one go about invalidation? Take, as an example, Ohm's law: $V = I \times R$ (volts = current x resistance). It is easy, with hindsight, to sketch the dangers it poses in a broader context: the pollution that generators create, the electrocution of wildlife, the harmful disposal of storage batteries, and so on. Yet even Ohm might have—*should* have—predicted some of it.[9] In this way, he would have invalidated his own theory in a global system.[10]

This being true, the notion of falsifiability must change, too. At present, we assume that scientific theories are verified as if they were true, with the possibility of falsification. But this is falsification within closed systems—systems with well-defined limits, which do not include everything.

All valid theories should not merely be falsifiable in closed systems but in an open or global system, in the context of all the relations in the

---

8. Paul Feyerabend wrote that science is "superior only for those who have already decided in favor of a certain ideology, or who have accepted it without ever having examined its adavantages and its limits" (Feyerabend, *Against Method*, 15).

9. It seems that Ohm did *not* predict how his law might affect the world. He was a theoretician, not a visionary. This was part of the reason for his initial obscurity (O'Connor and Robertson, "Georg Simon Ohm," third last para.).

10. In fact, the countless scientists who succeeded Ohm and made use of his theory ought to have *continued* to seek to invalidate his theory in a global system. Here, it is important to consider what would constitute invalidation. The effects of the *generation* of electricity have since been catastrophic, while, say, the electrocution of a relatively small number of people is presumably inconsequential.

world. In terms of a philosophy of relations, this will include ethics, even esthetics, because ethics and esthetics are derived from relations between things. The tightly woven relations of scientific theory need to be opened up to all the relations in the world—as far as this may be possible.

So as not to cause any confusion, we shall call the falsification we speak of *invalidation*. If a theory is not open to *invalidation*, then the scientific method is incomplete.

Scientific Progress

A failure to understand the "double isolation" of science may sow confusion as to how science truly advances. The traditional view is that, by and large, science advances through an *inductive* process—that is, by making broad generalizations from specific observations.

Yet consider that these specific observations have already followed the procedure of screening things out. That is, science has already minimized unwanted influences on independent variables—and, with this, has reduced the amount of thought that it gives to the world. It has excluded a great many possible relations in order to trace the relations that it does.

For this reason, the inductive process is inherently trammeled. Before one even puts it to use, it may have screened out the world.[11]

I draw on my own experience. In 2005, I designed the first of a new class of metal detectors. I named the principle CCO, for coil coupled operation.[12] I was already familiar with a TCO, or transformer coupled oscillator. This is governed by theory that, even in its full complexity, has little or no interest in outside influences on the inductance of the coils.

Now consider that the inductance of the coils and the frequency of the TCO may be influenced by coins beneath the soil—which are outside influences. In order to turn this into a metal detector, my mind needed to step outside of straight TCO theory to discover a principle that rested precisely on those influences that were excluded.

Science, therefore, requires not only the inductive method but something far broader—namely, intuition.[13] Albert Einstein wrote, "Knowledge

---

11. In some areas of life, it may not seem to matter too much that one screens things out—say, in writing a novel or staging an event. In science, however, it may be critical.

12. Scarborough, "CCO Metal Detector."

13. Mel Thompson puts it like this: "Does science not sometimes require imaginative leaps beyond evidence, in the formation of new paradigms within which detailed work and calculation can subsequently find its place?" (Thompson, *Understand*

is limited to all we now know and understand, while imagination embraces the entire world, and all there ever will be to know and understand."[14]

This has important implications for the scientific method. The inductive process should be taught as only one possible means of doing science or as one part of it, particularly in the early phases of the method. Further, the emphasis should be on a more imaginitive and frenetic thinking, which has its source in a broadly based education.

This is borne out by the fact that many notable scientists were interdisciplinary or multidisciplinary in their pursuits—among them, Archimedes, Leonardo da Vinci, and Albert Einstein. A notable example, too, is that of Yannas and Burke in chapter 26.[15]

Scientific Revolutions

Our observations may have a bearing on the theory of scientific revolutions.[16] By and large, in the process of screening things out, scientists' thinking is imprisoned by the scientific method—which is the normal condition of science. But a paradigm shift requires an eye for the wider canvas of relations between things, where one begins to discern something new through the *whole* of things.

Now consider that, as a science advances within a given paradigm, the work of scientific control is gradually done. No longer is there such an intense interest in what we should be screening out.[17] The liberty to think more broadly and creatively now becomes stronger. It stands to reason, then, that paradigm shifts occur at points of (by and large) maturity and stability—which in fact they do—where the imagination has the freedom to roam again.

Therefore, rather than paradigm shifts occurring through an accumulation of evidence until, so to speak, the dam bursts, they *fail* to occur until the constraints of the scientific method are weakened.

Philosophy, 127). The mathematician J. Henri Poincaré wrote, "It is by logic we prove, it is by intuition that we invent" (Webb, *How Numbers Work*, 105).

14. Porst and Sadeh, "Imagination-Based Interventions," 529.

15. This book was born in part through interdisciplinary postgraduate work.

16. To put it another way, the theory of science has a bearing on the history of science.

17. Thomas Kuhn called this *normal science*, which means "research firmly based upon one or more past scientific achievements supplying the foundation for its further practice" (Kuhn, *Structure of Scientific Revolutions*, 23).

The mathematician Laurent-Moïse Schwartz observed that there are mathematicians who do the groundwork, then "once their task is accomplished, the ideas of the scientists with a penchant for generality come into play."[18] While Schwartz may not have been thinking of paradigm shifts, this would suggest that relative dormancy in the scientific method is followed by special creativity.

18. Quoted in Barrow, *New Theories of Everything*, 223.

Chapter 30

Scientific Hubris

Our greatest pride is our downfall. Physical science, in its very nature, is the cause of civilization-wide crisis.

WE DEFINE *HUBRIS* AS excessive pride or self-confidence. Ancient Greek thinkers saw it as losing one's grip on reality, so leading, ultimately, to disaster. Many mythical figures were punished for it: Arachne, Icarus, Phaethon, Salmoneus, Niobe, Cassiopeia, and Tereus, among others. In the Hebrew tradition, "Pride precedes a disaster."[1]

Today, we experience a civilization-wide crisis, which includes, among other things, climate change, loss of species, new pandemics, diminishing fresh water supplies, social instability, economic jeopardy, and a good deal more. We all feel sure that something has gone peculiarly wrong—but what?

We are certain that this has much to do with ourselves—yet there is some confusion as to why this should be. We may list our rising population, pollution, consumerism, a lack of political will—among other possible explanations. The reasons may be *philosophical*, too: our increasing distance from nature, our neglect of the human sciences, our irrational optimism, or our short-term thinking.

Which, then, will it be?

In terms of a philosophy of relations, it has to be *all* of the above and more. Thomas Berry, a cultural historian and historian of the Earth,

---

1. Prov 16:18.

wrote, "Every reality of the universe is intimately present to every other reality of the universe, and finds its fulfillment in this mutual presence."[2] The problem must lie, therefore, in everything—in the food that we eat, the clothes that we wear, the devices that we use, the journeys that we make—in fact, almost anything we can think of. All are intimately interwoven with the problem.[3]

Physical Science

However, a single suspect looms large in the mix. It is the one thing, too, in which we pride ourselves most—namely, physical science.

Someone might object, "One cannot speak of physical science as *hubris*." Surely, science is humility incarnate. The journalist Douglas Preston wrote, "Hubris and science are incompatible."[4] Science is amenable to reason, it is free of opinion and bias, it requires mutual cooperation, it compels us to confront our own fallibility, and it liberates us from myths and servitude.

Yet this is deceptive. Superficially, we find it in the way that we *speak* about science on a daily basis. Science is not merely a method or a type of knowledge, but we frequently speak about it in terms we normally reserve for gods. It is "powerful," "eternal," "beautiful"—and scientists' own praise for science is no less effusive. The biologist Louis Pasteur wrote, "It is the torch which illuminates the world."[5] This—or something like it—has been the general view.[6]

However, in terms of a philosophy of relations, physical science (again, we shall say "science" for short)—for all its benefits—is a Trojan horse. It has destruction built in.

---

2. Berry, *Dream of the Earth*, 106.

3. We merely need to look across the nearest work desk to find numerous items that are intimately involved in diverse problems of the planet.

4. Lapchak and Boitano, "Reflections on Neuroprotection Research," 3.

5. Bhalerao, "Interacting with Dr. R. Chidambaram," caption.

6. In a sense, we have absolutized science.

> Unless scientific progress is balanced by another kind of enquiry, it will inevitably become an instrument of self-destruction.
>
> —JACOB NEEDLEMAN AND DAVID APPELBAUM[7]

Science, through its blindered focus on the scientific method, breaks the first ethical maxim of this book:

- *Think expansively and holistically.*

It breaks the second maxim, too:

- *Embrace those relations that exist in the world.*[8]

Science, by its very nature, is exclusionary.[9] x does not include not-x, y does not include not-y, and so on.[10]

A picture has stayed in my mind. In my youth, I lived on islands in the Pacific, where the bomb had been tested to the north and to the east just a few years before. One such test sought to determine the effects of nuclear weapons on ships. In Test Able, the bomb sank five out of seventy-eight vessels; fourteen were seriously damaged, and all were seriously contaminated.[11]

Shortly after the explosion, marines were ordered on board the remaining vessels to scrub the decks, to decontaminate them. But something was not understood. Scientists had failed to predict—or fully predict—the nature of radioactive contamination. The marines were exposed to high doses of radiation, and the ships were abandoned and sunk.

---

7. Needleman and Appelbaum, *Real Philosophy*, 12.

8. While none of our acitivities is whole, in the sense of being fully balanced, science *deliberately and radically* breaks these ethical maxims.

9. There is the argument that science excludes things that are so trivial that they can be completely ignored—for example, exoplanets. We shall see, however, that trivial things can and do take on great importance.

10. This reveals a further feature of relations as a whole. Because the relations in this world are boundless, and all things are related to all things, no matter what we do within a given region of relations, we cannot ultimately isolate it from the whole. The whole has priority over the parts.

11. Cowan et al., "Operation Crossroads," para. 5.

Scientific Hubris 213

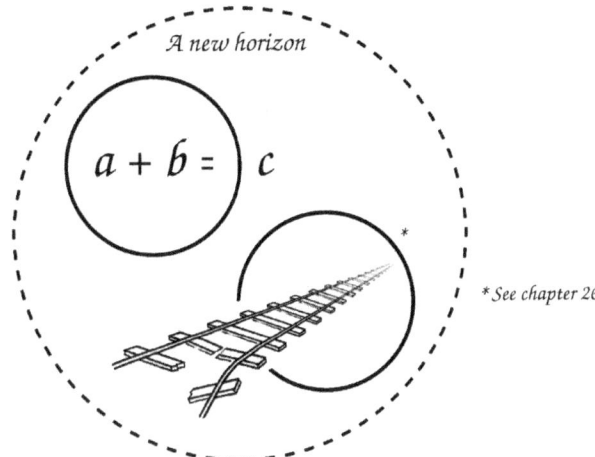

Science has often been unable to see—or to *foresee*—the damage that is done through those things that it has *screened out*. This has led various thinkers to suppose that science is responsible for the ruination of our world. Kamlah and Lorenzen wrote that science carries with it the "heavy price" of unpleasant and unforeseen side effects.[12] This, they wrote, is the most plaguing problematic of the scientific method.

In many ways, and at various times, the *consequences* of our scientific pursuits have been ill considered and poorly understood—and *continue* to be poorly understood—such that, in the words of Stephen Hawking, humanity is likely to "score an own goal."[13] We are in a situation today, wrote innovation journalist Luca de Biase, where "the scientific method must take into account the consequences of research." The problem, he wrote, lies in the method itself.[14]

In terms of a philosophy of relations, the scientific method has radically and deliberately reduced the world with which we work, so that unpleasant and unforeseen side effects are inevitable. This is compounded by the ease and speed with which these side effects are spread throughout

---

12. Kamlah and Lorenzen, *Logical Propaedeutic*, 127. Without the scientific method, we might invoke aetiological myths or practice senseless atrocities. This, too, carries with it a heavy price. In all cases, however, reason is short-circuited.

13. Shukman, "Hawking."

14. Biase, "Science of Consequences," 499–502. The problem lies, too, in the limited time we have to solve the problems. The physicist Max Tegmark observes that "learning from mistakes is not a desirable strategy" (Tegmark, "Wisdom Race," 205).

the world. Without our knowing the consequences of the proliferation of technologies, such consequences are multiplied among us.

> ### Excursus
>
> But is it truly *science* that is at fault? Do we not confuse science with the scientific method, engineering, or technology?
>
> Wilhelm Kamlah and Paul Lorenzen blamed "the very progress of knowledge itself."[15] While the careless or misguided *application* of science will in many cases be at fault, the bigger problem lies with the body of knowledge we accumulate through the scientific method. It is this *knowledge* that is deficient.
>
> Science proceeds by stripping away relations from the whole. It screens out unwanted influences on independent variables. With childlike optimism, it proceeds with far too little thought for the world.[16] It does so, too, on a vast scale. In every city and nation, day after day, it inexorably and powerfully imposes a reduced structure on nature.

Economics and Commerce

The sciences in general are not exempt. Our problems become deeper as we apply a reduced thinking to various fields. Among them, *economics* looms large, together with commerce.

While we have been taught to think that economics has to do with the production, distribution, and consumption of goods and services, today, it has become intensely mathematical and therefore *abstract*. This is not merely at the level of government and business, but everyone is now integrated with the system and feeds the system. Everything, in the words of economist Niko Paech, is now connected with everything, through streams of people, goods, and data.[17]

---

15. Kamlah and Lorenzen, *Logical Propaedeutic*, 127.

16. Arthur Eddington wrote that the scientist "rules significances entirely outside his scope" (Eddington, *Nature of Physical World*, 109).

17. Paech, "Destructive Dream of Progress," para. 5. At the time of writing, I

Scientific Hubris

Around the beginning of the twenty-first century, science, in general, entered a major new phase. Since then, it has two things at its disposal, which it did not have before: enormous processing power and vast amounts of data—or, to put it another way, extremely large data sets (big data), which may be analyzed computationally.[18]

It is hard to overemphasize the importance of this shift. Pedro Domingos wrote, "If every algorithm (a set of computer instructions) suddenly stopped working, it would be the end of the world as we know it."[19]

The goal of such algorithms, to put it too simply, is to *optimize* things—that is, to make things *fit*. Production must fit consumption, goods and services must fit our needs, foodstuffs must match our tastes, and so on. In the process, we strive for less wastage, smaller margins of error, and faster turnover. With this in mind, economics may more accurately be described as the *science of scarcity*.

There is no end to this tendency towards a better fit. The more data we have, the more we can tailor any number of things to our needs and desires: insurances, medicines, fashions, regulations, news, travel, and so on.

While the computational analysis of big data has many proven benefits, the algorithms are interested only in those things—*of course*—that the algorithms include. Most things are excluded. Say, the environment. It is no coincidence that as data proliferates and as our processing power increases, so do our global travails.[20] More and more—unimaginably more—is excluded from our thinking, and the "thinking" of the machines we have programmed to think on our behalf.

To put it another way, there are *externalities*—often, *global* externalities. The economist Henry Sidgwick called these "spillover effects,"[21] even before the full implications of this were understood. Oceans are polluted, glaciers retreat, bees are poisoned, toads go blind, people suffocate—in

---

bought a 100-gram packet of nuts. It was the product of eleven countries on five continents. Consider the airplanes, ships, and trucks that were needed to coordinate this, for my pleasure.

18. Pedro Domingos wrote, "Machine learning is the scientific method on steroids" (Domingos, *Master Algorithm*, 13). We have now been on steroids for nearly a generation.

19. Domingos, *Master Algorithm*, 282.

20. According to political reporter Lydia DePillis, if one fragments delivery van deliveries in the interests of a better fit, "in the worst-case scenario . . . the carbon emissions can be as much as 35 times greater" (DePillis et al., "America's Addiction," para. 10).

21. Medema, *Hesitant Hand*, 35.

fact, thousands if not millions of things go wrong besides, without their being represented in the algorithms.[22]

Paul Tillich put it radically: "The fundamental virtues in the ethics of a capitalist society are economic efficiency, developed to the utmost degree of ruthless activity."[23]

The science of scarcity is a form of totalitarianism, in that it requires complete obedience to the need for a better fit. In the words of Mel Thompson, "It attempts to take on the role of God, giving an ultimate justification and valuation of everything."[24] Such totalitarianism contravenes the ethical maxim of chapter 8:

- *Avoid totalizing tendencies.*

It would be tempting to decide that we may contravene this maxim, then correct course as we learn of the problems. However, the consequences of our actions are now potentially so big and so far-reaching that we may not have this liberty. Besides, reality itself proves to us that, in spite of our seeking to limit the damage, our global situation continues to degenerate, in crucial areas.

Finally, there is something greatly disturbing about the vast amounts of data we now process. We are all involved in the problem. We all feed the system. It is I myself at whom I may be alarmed.[25]

---

22. One may further note that extreme efficiency is very vulnerable to social disturbances and environmental changes.

23. Manning, *Cambridge Companion to Paul Tillich*, 128.

24. Thompson, *Through Mud*, 248–49. Stephen Hawking said in 2014, "Artificial intelligence could be a real danger in the not too distant future" (Hawking, "Stephen Hawking Interview," 2:43). In fact, the danger is now.

25. The journalist Michael Levitt quotes author James B. MacKinnon, "You name [an environmental problem], consumerism drives it" (Levitt, "There Is an Environmental Impact," para. 7). Further, the fact that face-to-face interactions are often replaced with digital interactions changes our ethics, since ethics is the product of how we arrange the world in our minds.

Chapter 31

Pseudo Ethics

The exact sciences present us with continual novelty and advances, while the human sphere is a quagmire. This may drive us to a false ethics.

WE ARE NOW IN a position to examine a kind of ethics that, on the surface of it, seems plausible but is not. We shall call it *pseudo ethics*. It is born of the fact-value distinction, which we have discussed through several chapters.

We take it for granted today that it is impossible to derive an "ought" from an "is." We are unable, through fact, to arrive at any value. All over, Hume's (supposed) insight has taken hold. What began with an insight almost trivial—that an "is" cannot produce an "ought"—has profoundly influenced our civilization the world over.

The is-ought problem leaves us with two logical extremes: logical positivism, on the one hand, or ethical intuitionism, on the other. Either we cannot speak about values at all, or we must speak about them *in the face of* our severance from the facts. Between these two extremes there is, so to speak, no place to land.

At this point—subtly, as if by stealth—we fill the value vacuum with *pseudo* values, borrowed from the realm of fact. Philosophically, this cannot be done—it is philosophical foolishness—yet it *is* done, and it happens like this:

It is one thing, we have said, to trace relations in the world around us, quite another to know how *others* trace relations in this world. While our physical world is more or less open to view, not so with worlds that exist inside other people's minds. Other people, who further hide behind semiotic codes: the raising of an eyebrow, for instance, a laugh, or an utterance. Like the chess novice who inserts a move into the grand master's game, as soon as we introduce *others* into the picture, there is a quantum leap of complexity.

This is not merely a practical problem. We have a problem in *principle*. By combining our own arrangement of things with the arrangements of countless others, we are combining vastly complex systems, with often unfathomable results. Small wonder, then, that we find it easier to speak about our world in factual terms rather than in human terms.

Further, in the human sphere, we often experience reversals and uncertainties—war, famine, and disease, among other things—while the natural sciences present us with continual novelty and advances. In comparison with the factual sphere, the human sphere is a quagmire. This leads us to a spontaneous privileging of the natural sciences.

Things Indicated

With a privileging of the natural sciences, then, we come to see those sciences as *indicating* value, where strictly they do not—and cannot. That is, we consider that they give us direction as to how we should *behave*. And so, clinical indicators determine our responses to a clinical situation,[1] economic indicators determine our responses to the economy, environmental indicators determine our responses to the environment, and so on. We automatically, impulsively, instinctively react to graphs, charts, and statistics, imagining that they give us not only guidance but criteria by which to act.[2]

This is illusory. While the natural sciences might *seem* to point us somewhere, strictly they do not and cannot. This is fact seeking to show us value. This is bridging the unbridgeable fact-value divide.

Jean-François Lyotard proposed that *efficiency*, above all, provides us with legitimation for human action today.[3] If we can only do something

---

1. The very reference to a patient as a "clinical situation" reflects the privileging of the natural sciences.

2. Jean-Francois Lyotard spoke of the "operativity criterion" and "technological criterion" (Seidman, *Postmodern Turn*, 28).

3. Lyotard, *Postmodern Condition*, 44. Max Weber similarly wrote about *rationalization*, through which wealth is presented as the highest good (Weber, "Science as a Vocation," in *From Max Weber*, 139).

more efficiently—or more profitably—we have a good enough reason to do it. In fact, all of society, he wrote, has become a system that must aim for efficient functioning, to the exclusion of its less efficient elements.[4]

Thus we calculate, tabulate, and assess things on the basis of "accounting calculations." Such calculations, writes sociology professor James Aho, have "under the banner of efficiency come to colonize themselves in virtually every institutional realm of modern society."[5] This is and has to be a philosophical mistake.

> Reason itself appears insane as the world
> acquires systematic totality.
>
> —THEODOR ADORNO

Not only this, but the idea of efficiency has introduced us to a life that many of us would not have imagined as children. We call it "increased objectification." We reduce organic fields of interest such as farming, building, nursing, even sports, to something increasingly resembling paint-by-numbers—while others become mere objects and instruments in our lives—delivering goods, taking samples, routing calls, and so on. We are processed, and we process others, on the basis of data sets.

We should not underestimate the extent of the problem this causes. With the advance of efficiency as a motive for action, we find widespread alienation today: feelings of boredom, powerlessness, normlessness, meaninglessness, social isolation, and despair, which often did not exist in earlier times. The philosophical economist Karl Marx considered that we have been overtaken by "commodity fetishism,"[6] where we devalue the human sphere as we overvalue *things*. The theologian Samuel Goodwin wrote, "We are just a number."[7]

Through pseudo values, borrowed from the realm of fact, we are dehumanized. In fact, this must be the case as long as we take numerate

---

4. Herbert Marcuse pointed out that this is promoted by modern education, which is designed to make individuals useful in the production of goods and services (Marcuse, "Individual in Great Society," 34).

5. Aho, *Confession and Bookkeeping*, xi.

6. Marx, *Capital*, vol. 1, §4.

7. Goodwin, *Just How Dumb*, 71.

approaches to human affairs. Cold fact invades the complex, subtle relations represented by the human sciences—in fact, by life as we live it.

Of course, efficiency has its *positive* aspects. We receive efficient service, design an efficient machine, or have an efficient economy. This alone raises the status of efficiency in our thinking. However, where efficiency represents legitimation for human action, it has no proper place.

My own maternal grandfather provides an analogy. Having been invited to play a duet on musical instruments, he performed harmoniously, flawlessly, until he neared the end. Then he speeded up over his fellow musician and declared with laughter, "I win! I finished first!" His partner protested: "It's not *about* that! It's about harmony, beauty—*music!*"

From one point of view, my grandfather's efficient finish placed him first. From another point of view, he seemed to miss the point of a musical performance altogether.

Day by day, there are things we value more highly than efficiency. We know it when we stop for a cup of coffee at a street café, take time out with friends, or go on a sightseeing tour—even take part in a funeral procession or witness a state investiture. Some of life's most precious moments have little or nothing to do with efficiency or profitability.[8]

Science Demoted

Due to our spontaneous privileging of the natural sciences, we raise them above and separate them from the *human* sciences. We see this in the most basic features of our educational systems today.

It begins in our junior school years, where the human and the natural sciences are allocated separate periods, separate books, and, often, separate teachers. Such separation continues through our high school years and, finally, through our university education, where it is written into the very ground plan of our campuses: the physics building over here, the humanities building over there.

Not only does such thinking apply to our university campuses. It has worked its way through the entire fabric of our (post) modern society. Whether it be time slots in television news shows, storefronts in a shopping mall, or portfolios in a government cabinet, all reveal, to

---

8. The Critical Theory of the Frankfurt School highlighted the neglect of cultural and personal values, which they deemed had fallen prey to market forces (Mautner, "Frankfurt School," in *PDP*, 207–8).

Pseudo Ethics

some extent, the separation of the natural and the human sciences. Karen Barad summed it up in a word: "bicameralism."[9]

To put it most simply, we separate things from things. We create *classes* of things, which tend to be mutually exclusive. We separate windows and cabbages and football players from protons and plutonium and DNA—and we separate both of these from mathematical objects besides.

An important aside suggests itself here. In previous chapters, we noted that, when we do science, we must minimize unwanted influences on independent variables. This means that we must radically reduce the full range of relations on which we otherwise draw to function holistically.

Physical science, by its very nature, is strictly limited in scope. We may go so far as to say that it is one of *Homo sapiens*'s more primitive pursuits. For all its complexity and sophistication, it deals only with the simplest of relations between things, namely, those that we can quantify. At the same time, it is the human sciences—the *uncertain* sciences so-called—that deal with more complex, wide-ranging relations, and global systems. These present the greater and more important challenge for human knowledge and wisdom.[10]

The privileging of the natural sciences represents a fundamental distortion of values.[11] Like a stricken cruise liner, the all-important passenger decks roll under the water, while the machinery turns upward. In fact, our scientific knowledge will necessarily run ahead of our social competence, until we develop an ethics that is more broadly based on relations, rather than a pseudo ethics through which we grasp at the ghost of an ethical basis in facts.

---

9. Barad, *Meeting the Universe Halfway*, 58.

10. Ludwig Wittgenstein wrote, "We feel that even when all possible scientific questions have been answered, the problems of life remain completely untouched" (Wittgenstein, *In Search of Meaning*, 40).

11. Karen Barad asks, "Are there ... openings that exist within physics that trouble its hegemony, its authority, its unapologetic epistemological imperialism that claims to cover all of space, time, and matter?" (Barad, "Re-membering Time," 36:12). The answer here is yes.

Part V

Classics

The fifth part of this book addresses some of the more classic themes of metaphysics, seeking to integrate time-honored subjects with a philosophy of relations. But first, we need to drive deeper with the *foundations* of metaphysics.

Chapter 32

Deeper Relations

The traditional scheme of things distinguishes between things and relations. We now need to drive deeper. We need to eliminate "things."

IN THE COURSE OF this book, we have made various observations about things and relations. These have helped us explain failures of past philosophy, develop a new moral philosophy, address the fact-value distinction, and bring into harmony the various "languages" that we speak. However, our concepts have now come as far as they will carry us. For some of the subjects that lie ahead—among them, the mind-body problem, free will, and history—we need to drive deeper and focus more intently.

In chapter 7, we sought to describe the *number* of things and the number of relations beween them. We noted there—in the example of the heat that causes the thunder—that things and relations may be stranger than they seem.

When we seek to define things and the relations between them, we find an infinite regress. When we examine *relations*, we find new *things* in their place—and where we find new things in their place, we find new relations.

We may imagine a fractal image. The more closely we look at it, the more we see—and more and more, as we travel ever more deeply into the image. Where, then, does a fractal image end? It never does. The same it is when we look more closely at things and relations. Their essence flees away.

David Hume, we said, recognized that we could not, by merely studying the relations between things (he called them "objects"), discover how they *ought* to be related.[1] On this basis, he separated fact from value—fact being the way that things *are* related to each other, and value being the way that they *ought* to be.

Many philosophers adopted a similar scheme—a scheme that, in the case of Hume and Wittgenstein in particular, led them to believe that the door to a rational ethics was closed.

Consider now what happens when we compare this received view of things and relations to a view that we fleetingly put forward in the opening chapters.

In chapter 2, Mel Thompson described our world as "a seamless web of causality that goes forwards and backward in time and outwards in space." "Seamless"—if we take this to mean exactly what it says—implies that there is no separation or individuation of things.

This was not the view of past metaphysics and philosophies, which were *full* of seams—between things, objects, entities, elements, parts, events, actions—each of which had some kind of outline and was separated from the rest. In such a view, we said in chapter 3, things *cohere* in space and time: a pot coheres, a person coheres, a planet coheres.

Bacon's Evil

In chapter 4, we described a *subatomic* world of words. We said that hundreds of thousands of words in our language contain a world of relations inside, which may bind with relations inside other words, anywhere and everywhere.[2]

Francis Bacon wrote that there is an "evil" in dealing with natural and material things. As soon as we seek to define such things, "definitions themselves consist of words, and those words beget others."[3] That is, there are words inside words, and inside those words are words, presumably to infinity.

---

1. Hume's terms may be confusing: objects and relations on the one hand, and matters of fact and relations of ideas on the other. The two are not the same.
2. Again, we do not strictly find relations "inside" words or "outside" words, or anything else for that matter. It is for the sake of a working model of the word that we say that words contain a world of relations inside.
3. Bacon, *New Organon*, §59.

Deeper Relations

This does not mean that a language contains an infinity of words but that the words that define words, and the words that define the definitions in turn—and so on—are as good as infinite.

The novelist Hermann Hesse likened it to a man who researches a subject in a library. One book leads to another, which leads to another, and another, until his search reaches into the whole library.[4] It seems much like having your money in a bank, which has its money in another bank, which has its money in another bank, and so on. It is not hard to see that the money will never be found.[5]

> Philosophers of language typically talk about "objects" . . . as physical givens when they are really talking about perceptual signs.
>
> —CHARLES OSGOOD

Here we have the familiar scheme of things and relations—yet now we see things within things, and relations within relations. Where, then, does it all end? It ends, in fact, nowhere. It all ends in nothing—or everything.

The things that we thought to be things dissipate in the endless expanse. The relations that we thought to be relations vanish into (almost) infinity. We are left neither with things nor relations. In the final analysis, our words are some kind of nothings—or, to be more exact, their *referents* are. We shall call such words "dissipated words." They *dissipate* in the endless expanse.

In fact, it is interesting to contemplate just how far each word might reach into our world. On the surface of it, nearly every word we speak may conjure up a whole universe of associations. Using a word seems like striking a gong, which causes the whole universe to reverberate.

---

4. H. Hesse, *Journey to the East*.

5. Lao Tzu wrote, "Reaching and not touching it, we call it ethereal" (Lao Tzu, *Tao te Ching*, viii). Lord Anthony Quinton noted that "to possess a concept is not to own some easily identifiable article" (Quinton, "Concept").

Excursus

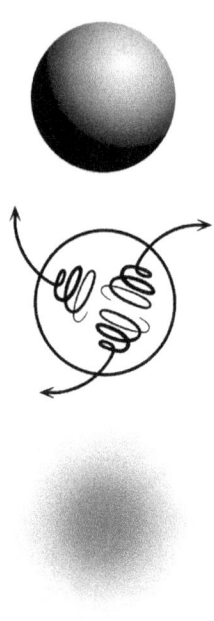

How, then, does this fit with the concept of a word, as previously described? A word, we said, contains a world of relations inside, which may bind with similar relations everywhere. While we do not now discard that model, it is simplistic.

In the image at top left, the original, atomic word is the familiar view.

Beneath it, the *subatomic* word is the model we introduced in chapter 5.

At the bottom, the *dissipated* word is similar to the subatomic word—however, it now offers us a more subtle picture.

Not only do the relations inside a word combine with the relations inside other words. Words now *dissipate* in the endless expanse—and so, in fact, do things. Entities, wrote the philosopher Achille Varzi, are not sharply bounded. "On closer inspection, the spatial boundaries of physical objects are imaginary entities."[6]

Here we reach an important and necessary conclusion. The way that we use words, *upon analysis*, proves to us that we do not treat them as referring to *entities*—or to things that have a distinct and independent existence. Recall that things are the contents of words. Therefore, things are *like* words.

In other words, when we refer to *things* so-called, we do so only by way of shorthand. The identity of those things does not lie in themselves. It is always elusive, always deferred.

Imagine that I hold a drop of red ink, with its sharply defined contours—hanging tremulously from an applicator over a glass of water. I

---

6. Varzi, "Boundary," 1.2. Percy W. Bridgman wrote that "we see that an object with identity is an abstraction corresponding exactly to nothing in nature" (Bridgman, *Modern Physics*, 35).

let it drop. When it hits the water, it explodes into a cloud of color in the water, which eventually fills the whole glass. Similarly, we habitually treat words—and things—as if they were the drop of ink on the applicator, while, in reality, they resemble the cloud of ink in the water.[7]

Take the simple example of a *car*. A car is, according to the *Oxford English Dictionary*, "a four-wheeled road vehicle that is powered by an engine and is able to carry a small number of people." Four, in turn, is "a cardinal number, equivalent to the product of two and two; one more than three, or six less than ten." A wheel is "a circular object that revolves on an axle and is fixed below a vehicle or other object to enable it to move easily over the ground." A road is "a wide way leading from one place to another, especially one with a specially prepared surface which vehicles can use." A vehicle is "a thing used for transporting people or goods, especially on land, such as a car, lorry, or cart." And so on.

We are hardly a quarter of the way through the definition of a car, and we have heaped up new words: object, product, axle, surface, revolve, fix, enable, power, circular, small, cardinal, prepared, and so on—words that themselves wait for their own definitions.[8]

In this example, too, we see that *things* are just as complicated as words. We say of something, "That's a thing," yet it really cannot be that thing if we remove the axle, the surface, or the power, and so on.[9]

The multiplication of words is only possible because our reality is fit for it—or, to put it another way, is equally (call it) *diffuse*. We suggested it as early as chapter 3. Things—although they may seem to cohere—do not. Again, in the words of Achille Varzi, the boundaries of objects are *imaginary*.[10] However, our scientific study of boundaries is not as far advanced as our scientific study of words.

The content of dissipated *words*, therefore, is dissipated *things*. It is not as though we use dissipated words to speak of concrete things—as

---

7. Even where our words are carefully defined, so cutting off—or so we imagine—most of their possible relations with the world, we find that much the same is true, namely, that their essence *dissipates*.

8. Norman L. Thomas observes that "the definitions we find in the dictionary are circular" (N. Thomas, *Modern Logic*, 191). But not strictly circular—rather, without end.

9. This applies both in the simplest way and more theoretically. If our universe did not contain any of these things, there would be no car.

10. Varzi, "Boundary," 1.2.

if the definitions of words reach into all of reality, while the essence of things does not. Things, too, reach into the whole universe.[11]

Therefore, there is now room to entertain thoughts of *dissipated* things. This is a low-level idea that will soon become critical to causality, free will and history, even the existence (or not) of God.

Nothings

We live with a strange paradox, then. Intuitively, we believe in *things* and speak about things, yet our very own words and thoughts prove to us that they reach into an endless expanse—or into the entire library, as it were.[12] There is everything yet nothing we can *fix* on. Ultimately, there are no things; there are no objects, no entities, no events—therefore, there are no facts.

To put it another way, that which happens in the vast neural network that is or contains the mind does not take the same form as things that exist—or are thought to exist—in the world. A complete *translation* takes place, in order for them to enter the mind,[13] and in the mind, there are no concrete objects. "A map is not the territory," wrote the semanticist Alfred Korzybski.[14] We should not confuse the two.

---

11. In the evolution of human thought, we have, in important ways, paid more attention to the arrangement of words than the arrangement of things. Above all, we have thought a lot about the boundaries of words (which is their meanings), but not a lot about the boundaries of *things*. While the science of the boundaries of words is far advanced, the science of the boundaries of things is in its infancy.

12. The same may be said of equations, which always point to other equations. For example, Ohm's law points to the power rule, and to Faraday's law, and so on.

13. The editor of The Penguin Dictionary of Philosophy, Fritz Mauthner, wrote that language is a veil which prevents access to reality (Hermans, *Interbellum Literature*, 212). Max Black noted, "It is not easy to understand how [language] can be learned or embodied in a neural mechanism. . . . The secret seems to reside in something no less fundamental than the apprehension of relationships in general" (Black, *Labyrinth of Language*, 66).

14. Korzybski, *Science and Sanity*, 58.

Deeper Relations

> **Excursus**
>
> No things? No relations? No facts? If this really were the case, we could dismiss this philosophy as baseless. One cannot describe a philosophy on the basis of *nothing*.
>
> However, when we speak of dissipated words—and dissipated things—we speak on a *theoretical* level. One might say the same about, say, nuclear physics. How can we continue to live in a world that we now know to be, as someone put it, "all space, and none of it empty"?
>
> Of course, we say that this is the *theoretical* view—and mostly, it does not concern us as we go about our everyday life. So it is with dissipated words versus atomic or subatomic words, in the earlier parts of this book. Our more commonsensical views are useful, but for certain insights, we need the theoretical view.

We didn't stop drinking water when we discovered that it was made of two gases, namely, hydrogen and oxygen. Yet our new understanding of water changed everything. So it is here. The scheme we have used in previous parts of this book—in particular, the great things-relations distinction—is now temporarily suspended.

We need now to emphasize—and cannot emphasize enough—that, as we enter the following chapters, we are entering upon a strange new world of *theory*. Let those who enter here be aware that we are now shifting gears: ☡.[15]

---

15. The Nicolas Bourbaki group use a "dangerous bend" sign to mark passages that are tricky on a first reading or contain an especially difficult argument (Strehl, "U+2621 Caution Sign").

Chapter 33

Causal Relations

The concept of free will has everything to do with cause and effect. Cause and effect, in turn, is a form of value judgment.

No matter what we may think about it, we know that there is a problem. We know that things are physically determined. I line up dominoes in a row and topple the first of them with my finger. It is certain that the whole row of dominoes will fall.

Am I then subject to the same kind of determinism? Or perhaps, to what *extent* am I subject to such determinism? Is it possible for me to escape my own inner person? My own circumstances? My own history? To what extent can I shape my own future? Am I free to choose my own thoughts? My own actions? Am I free to believe? Each of these questions would seem to present us with a range of mightily confusing answers.

Free will has everything to do with cause and effect—and once we understand the dynamics of cause and effect, we should be in a better position to understand free will.

In fact, this takes us back to the issue of fact and value. We said that we may judge that the relations that we find in this world are as they should be, or we may judge that they are not. No matter whether we are dealing with "is" or "ought," both may be stated in terms of "ought," or value.

Now take our value judgments, and add the factor *time*. We may judge that the relations that we find in this world—*through time*—are as they should be or are not. Therefore, cause and effect, which is (usually) a sequence in time, is a value judgment.[1] It is a value judgment, because it judges how things ought to be arranged in this world—rather than the way they are.

Some will judge causes one way, some another. Consider the outbreak of war. This may have been caused by the generals—or to some, by poverty, pride, or human cruelty—to some, by drought or plague—to some, by hexes or ghouls. Or it may have been caused by God—in which case, we ought to jettison all other notions of causality.[2]

Whatever the case may be, when we judge what it is that caused a thing, we state, in effect, that this is how it *ought* to be or ought not to be. This is not to say that we manipulate the facts—although, sometimes, we do. Rather, we decide where the *limits* lie of the facts that we choose. We scope the system, as we said in chapter 26. We select particular facts as the essential ones.

Cause and Events

What is cause and effect? This calls for careful consideration.

In our everyday descriptions of our world, we say that event A caused event B, event D caused event E, and so. Simon Blackburn wrote, "Causation is the relation between two events. It holds when, given that one occurs, it produces, or brings forth, or necessitates the second."[3] The burrowing aardvark caused the dam to burst, the lightning strike caused the thatch to burn, the medicine caused the patient to rally, and so on. Of course, this is putting it too simply.

Yet more than this, we notice again, in Blackburn's definition of causality, the separation of events, on the one hand (things that happen), and relations, on the other. It is the distinction between *things and relations* that we have seen many times, in various forms, in this book:

- objects and relations (David Hume)
- objects and arrangements (Ludwig Wittgenstein)
- things and relations (Bertrand Russell)

---

1. Above all, states of affairs. If they do cause anything, they may not reveal a sequence in time.

2. The kinds of causal explanations that can be distinguished are deductive, probabilistic, causal, functional, purposive, teleological, and genetic.

3. Blackburn, "Causation," in *ODP*, 59–60.

- nouns and verbs (in language)
- variables and operators (in mathematics)[4]

The distinction between things and relations is, however, illusory. We passed beyond such thinking in the last chapter, when we ruled out the existence of things, objects, events, and the like—or rather, *self-contained* things, *invariant* things. All things and relations dissipate in the endless expanse.

Consider the following analogy. In front of my junior school class, I put ninety-nine marbles on a traditional scale. Then, when I drop the hundredth marble on the scale, it tips. I ask my class, "Which marble tipped the scale?" I know that they cannot know. They were too far away to see it.

One comes forward and says, "It was this one"; another, "That one." They are adamant, too, that they can prove it—and they do. They pick up this one, then that one, to show me that it tips the scale. Another says, "It was all the *red* ones"—and proves it, too.[5]

The analogy is limited in its usefulness yet illustrates the point that cause does not lie in a single event but in all events combined.[6] While we can identify them and prove them, every one, reality is really fooling us. Bertrand Russell wrote, "If the inference from cause to effect is to be indubitable, it seems that the cause can hardly stop short of the whole universe. So long as anything is left out, something may be left out which alters the expected result."[7]

---

4. One may add many related distinctions, among them snapshots and motion (Zeno of Elea), being and becoming (Plato), features and dispositions (Aristotle), ergon and energeia (Alexander von Humboldt)—static and dynamic, properties and relations, structure and process, and so on.

5. One may take the simplest example. I put a pencil to this page and draw a mark. On the surface of it, I caused the mark. Or was it something on the page? Or my own psychological impulses? Perhaps the pencil caused the mark? Or its graphite core? And so on.

6. Percy W. Bridgman wrote, "The causality concept is relative to the whole background of the system which contains the causally connected events" (Bridgman, *Modern Physics*, 91).

7. Russell, *Our Knowledge*, 229. If Bertrand Russell had taken this insight more seriously, he would surely not have espoused logical atomism as he did. The philosopher of science Karl Popper summed up *Hume's* thinking on this: "We can never know with 100 percent certainty that the event we are calling the cause is the 'true' cause.... Even designating where one event ends and the next begins is in many ways arbitrary or subjective (indeed, requires a theory)" (quoted in Alexander and Winne, *Educational*

There is a simple consequence to the interrelatedness of all things. It may seem too simple at first to mean anything—and yet it means everything. If all things are related to all things, and all things influence all things, then to isolate one thing and to call it a thing that *causes* something cannot strictly be correct.[8] In fact, in many cases, our isolation of causes has been disastrously one-sided: politically, philosophically, scientifically, and in many other ways.

There are no events A; there are no events B; nor are there Bs caused by As or As caused by Bs, because A and B—as with all things—strictly do not exist. There are, so to speak, neither variables nor operators.[9]

In order to establish cause and effect, we need to be able to *define* both A and B—but ultimately, such definition is beyond us.[10] Our *definiens* (which define things) reach into the endless expanse. When we define A and B, we remove much of the *synessence* (every possible aspect) of A and B. To put it another way, we reduce reality. We strip away a lot of what is truly going on. To the extent that we define a certain cause A, or a certain effect B, we proportionately exclude all other causes and effects.

While this might seem at first to stretch the imagination, it is a necessary conclusion: cause and effect cannot be shown to exist. This applies to every cause and effect, whether upward or downward, backward or forward, metaphysical, physical, or mental. Ernst Mach wrote simply, "There is no cause nor effect in nature."[11]

Psychology, afterword).

8. One speaks of causes that lie within my personal "event horizon." Many things do not.

9. The philosophical musician Jeffrey Werbock reportedly observed, "We think in terms of causality. It starts with perception. Perception is the act of converting the Unity of existence into the diversity we perceive, so we can perceive it. Once we perceive diversity, we begin to think causally." See http://www.mugham.net/articles for more observations by Werbock.

10. Definitions have a profound effect on our conception of cause and effect. These include our worldview—such that we think that one and the same behavior may be caused by (for example) sin or genes.

11. Mach, *Science of Mechanics*, 483.

> **Excursus**
>
> But does not cause and effect quite simply *work*? He lit the fire, I broke the urn, they split the atom. All such events, surely, reveal obvious cause and effect.
>
> It is important to note that our *experience* of cause and effect, and our *experience* of the things and relations on which they rest, is unchanged. Yet when we examine all these things closely, we see that none of them are real. At least, they are far from what we *thought* they were.
>
> By way of analogy, one may imagine a glass marble, which I tap on a tabletop. On first impression, it is hard and round. Yet as I peer into it with ever stronger microscopes, I first see its molecular structure, then its atomic structure—then I travel into the subatomic world that is mostly void. Yet even though I call it "mostly void," I still experience the marble.

Our analysis of causality leaves us without a reason to believe in cause and effect—even if we are naturally disposed to thinking that way.[12] In fact, any *empirical* debate about the existence of cause and effect—or of free will and determinism—is out of the question. It all rests on the presupposition that A and B exist.

Someone might object. Even if we have no things, no objects, no entities, no events, no relations (and so on), we still have a *reality* that is bound by the laws of the universe. There is, therefore, *some kind of something* that is not free.[13]

Yet every scientific law is about As and Bs. Whatever is out there, it has nothing in common with such a scheme—that we can know of.[14]

---

12. The traditional founder of the Samkhya achool of philosophy, Kapila, around the seventh century BC, reportedly criticized the concept of cause and effect. Rather, he saw effects as being preexistent in their causes (Ambedkar, "Kapila," in *Buddha and His Dhamma*, §2). Friedrich Nietzsche may have approached such a view when he wrote that cause and effect "is a question, not of succession, but of interpenetration" (Nietzsche, *Will to Power*, 336).

13. One might claim, too, that things exist regardless of any definitions—yet recall our observation that "a map is not the territory." We are using the map.

14. In this context, consider the evolutionary biologist Richard Dawkins. Things

Determinism can exist only if one employs a certain mode of thought, which is here eliminated.[15] It may be a new way of putting it, but it is not a new idea. Albert Einstein wrote that determinism is a feature of theories, rather than any aspect of the world directly.[16]

We see, then, that a single, simple idea—that As and Bs cohere—which has been accepted and handed down from generation to generation—has muddied our thinking in the crucial area of causality. It serves as another evidence of the importance of apparently inconsequential concepts which infiltrate our thinking everywhere, with potent effect.

We know, of course, that our beliefs about free will have a profound effect on us. Cause and effect is not merely something we observe impassively in the world around us.[17]

Wherever we attribute cause, there we lose our sense of freedom. Yet wherever and whenever we are able to change the relations that we trace, there we come to see causes differently—and see our world in a different way.

Consider a farmer who has suffered the failure of his rice crop. He blames it on the weather. But no, says the neighboring farmer. His circumstances are quite the same, and his own rice crop is thriving. "It's proverbial!" he says. "Blame it on the weather!" Yet notice that, by attributing blame, the first farmer admits to being bound in a cause-effect relationship of which he is not the controller. Yet his neighbor has no such sense of bondage. One farmer's freedom may be another farmer's bondage.

We all *believe* in things that cause things. Therefore, we all have a subjective experience as to whether we are controlled or not, and to what extent: by material entities, circumstances, or even our own selves. This profoundly affects us and the way we see our world. It makes the

---

do not occur at the level of meaning, he wrote, but at the level of "selfish molecules" (Dawkins, *Selfish Gene*, 7). To believe this, he needed to assume the existence of As and Bs. He was merely shifting the action arena, so to speak. This is the case with all ontological reduction.

15. A critical question is whether this opens up the possibility of causes beyond our world—or alternative causes within it. Among other things, the nonexistence of cause and effect as we know it may open the door to downward causation.

16. Fine, "Einstein, Albert." Compare the theoretical physicist David Bohm, who wrote that quantum theory is "the dropping of the notion of analysis of the world into relatively autonomous parts, separately existent but in interaction" (quoted in Capra, *Tao of Physics*, 150).

17. One might prefer to say that we create the *illusion* of cause and effect. However, things are concepts, and concepts are things. Our concepts are as real as it gets.

difference, too, between a free spirit, on the one hand, and a person who is repressed, inhibited, or compelled, on the other. We are as free as we think we are.[18]

Are we free, then, or are we determined? Or are we, perhaps, randomly triggered in various ways—say, through quantum indeterminacy?

Set aside again what we *do not* know, and consider what we *do*. We have established that, strictly speaking, there is no cause and effect—because there are no As and there are no Bs. These are illusory. We know, however, that we have the power to *impose* As and Bs on our reality.[19] We do it all the time.

Here is the conclusion. Since reality itself does not contain As and Bs, we ourselves create these from nothing—*ex nihilo*. With this, we see that, while cause and effect itself seems omnipotent, it is freely created in our own minds.[20] Outside of our own minds, it does not exist.[21]

I take a shot at a clay pigeon. It shatters in the sky. I have chosen my cause and my effect. I might just as well take a shot at a nearby church bell—and I do. This rouses an entire village from its afternoon slumbers. Here I have chosen a *different* cause and effect. While the cause has its effect, and while the cause and effect seem bound to each other by the laws of nature, in fact, I *create* them.

Of course, I cannot aim at the church bell and hope to hit the clay pigeon. I am not free to choose a cause without its effect. I choose the cause *with* the effect. In this lies free will.[22]

This works backwards, too. I may create the cause and effect of *past* events. To use the earlier example of the outbreak of war, I may choose any one of a number of causes for the war, and it may plausibly be correct.

---

18. This is close to the view of various existentialist philosophers, among them Martin Heidegger and Jean-Paul Sartre. The Thomas theorem states, "If men define situations as real, they are real in their consequences" (Chandler and Munday, "Thomas Theorem").

19. This is one of the core interests of new materialisms, which, according to social theorist Nicholas J. Fox, cut across foundational dichotomies such as "nature/culture, human/non-human, animate/inanimate, agency/structure, and mind/matter" (N. Fox and Alldred, *Sociology and New Materialism*, 13).

20. One could say that the causes and effects we invent are themselves caused in some way. Yet then, we would need to revert to a position that there are valid As and Bs, which there are not.

21. In other words, we include not only events that we ourselves cause but all events that we perceive as causes.

22. To see a cause without its effect is merely to see more narrowly.

It was the generals. No, it was the drought. Or, less plausibly, for some, it was a hex. No, God.

> ### Excursus
>
> Someone might say that we have truly lost touch wih reality. We create causality? From nothing? This is surely a complete departure from common sense.
>
> In fact, it is the opposite view that defies common sense—namely, that we must believe in the things and relations that form the backbone of causality. Simon Blackburn summed it up, "The idea that causation is a matter of discrete events joined by links is highly problematic."[23]
>
> In the early twentieth century, the philosopher Alfred J. Ayer pilloried Francis H. Bradley for holding similar views, and it stuck. Yet it was Ayer who, not long after, said that his own philosophy had been "all of it false."[24]

Big Concepts

Wherever we attribute cause—that is, wherever we judge that it *had* to be this way, where it did not—there we lose our sense of freedom. We are compelled by causes that we judge to be beyond our control.

Wherever *causality* walks onto the stage, there the "big concepts" of life lose their meaning. Francis Schaeffer noted that, in a closed system of cause and effect, love has no place, nor morality, nor freedom. Things become mere links in a causal chain.[25] Everything else is excluded. This leaves one with a barren view of reality.

---

23. Blackburn, "Causation," in *ODP*, 59–60.
24. Ayer, "A. J. Ayer," 34:10.
25. People, too, become mere elements of a system.

> Free agency is the power to decide according to our character.
> —CHARLES HODGE

As our lives become easier to predict, more manageable, and more secure—in other words, where cause and effect prevail—there we lose something of the *essence* of life: the wind in our hair, the smell of the rain, the chill of the evening, the soil under our feet, or the rising of the sun.[26]

In their place, we have climate control, artificial lighting, programmable cooking, and so much more, which so often represent a loss of accident and freedom in our lives. While we may *choose* such living, we were not "made" for it. The writer and philosopher Susan Sontag wrote, "All the conditions of modern life . . . conjoin to dull our sensory faculties."[27] There is much truth in that.

Further, when our lives are dominated by causality, we take past events to be the cause of our present situation—and present events to be the cause of our future situation. In so doing, we imagine that our present situation is held in the vicelike grip of circumstance and obligation or that our future situation is bound by fate or determinism. Alternatively, we justify a *calculated* approach to all that is to come—our health, our relationships, our personal safety—and so we become slaves to our own order.

Not only do we impose ideologies on others, so brutalizing them. We may brutalize our own selves, becoming the slaves of habit, principle, or tradition.

---

26. We lose, too, something of the *subtlety* of life. Thomas Huxley wrote that scientific progress means "the extension of the province of what we call matter and causation, and the concomitant gradual banishment from all regions of human thought of what we call spirit and spontaneity" (Huxley, *Physical Basis of Life*, 31).

27. Sontag, in "Forbes Quotes." An issue of freedom that may seem remote, but could become critical, is the habitability of our planet. We take for granted freedoms such as breathing freely, exploring our surroundings, choosing our food, and so on. Such freedoms could end, if the planet becomes less habitable.

Chapter 34

Historical Relations

History traces relations between things—as such, representing value judgments. Its truth is decided by independent, intelligent judgment.

CAUSE AND EFFECT IS not only important to the present. It is important to the past. It has everything to do with *history*—where history is a river of cause and effect. Thus our observations about the problems of causality apply to historical events. Not only this. In the last chapter, we went so far as to say that causality does not exist—that is, not in the world. This adds a new dimension to the problems of history.

We are all familiar, on a personal level, with various problems of history. We all revise the small histories that we ourselves create, with words such as "I misjudged him," "If only I had known," "I can't remember it for the life of me," or "I had forgotten how it was." Our own practice of history, therefore, is often subject to review and correction.

Much the same is true of history on a grander scale. History is never about certainties, and historians generally commit to no more than probabilities—even if the probabilities should seem overwhelming.

History, unlike the natural sciences, is unrepeatable. We cannot retrieve the facts if we missed them the first time. History, too, will always represent a loss of information—a reduction of the events themselves. In fact, no history can truly encompass all that happened in the past.

Even if we should select a limited and (seemingly) manageable part of history—say, regional history, military history, urban history, or family history—even evolutionary history, stellar history—this is not isolated from the whole. Any history, no matter how big or small, is ultimately a global system and will run up against the limits of completeness.[1] It will, so to speak, be frayed at the edges.

History is Relative

There is, however, a problem that overshadows all the rest. The facts, wrote the historian Charles A. Beard, are an act of choice, conviction, and interpretation regarding values.[2] Facts do not select *themselves* or force themselves into any fixed scheme of arrangement in the mind of the historian. Their selection and arrangement is subjective. Wherever we trace relations between things, we are selective and, in the process, de-selective.

This refers not only to a de-selection of facts through our own biases, prejudices, or negligence—deliberately de-selecting, say, proletarians or Blacks or traumatic events from our story.[3] There is a kind of de-selection where we are not so much as aware of the de-selections we make.[4]

As we write the story of progress, we may fail to notice the degradation of the environment. As we write the story of empire, we may fail to notice the destruction of social fabrics. As we write the story of enlightenment, we may fail to notice the deficiencies of our newfound knowledge.[5] It is called collective blindness.

Our simple observations about the law of noncontradiction come back into play. Every x casts away not-x; every y casts away not-y; every z casts away not-z. A is not not-A; B is not not-B.

But it is more subtle than this. If the world is one seamless whole, where everything influences everything else, then selecting and de-selecting things—in this case, events of history—is something we cannot

---

1. In the case of *religious* history, a global system may, in fact, help us solve the question of the trustworthiness of such history, as we shall see.

2. C. Beard, "Annual Address," 220.

3. In fact, many presuppositions of the past were highly rated, where now they are so unfashionable as to raise eyebrows. For instance, that history serves the purposes of the state, that it is about great men, or that it has a goal.

4. Insofar as it is possible to make de-selections one is not aware of. Some de-selections may, however, go back to earlier de-selections (say, the university I chose).

5. Stephen Toulmin held that systematization necessarily "involves omission" (Toulmin, *Cosmopolis*, 200).

strictly do.[6] History is about relations between events—which is things that happen—and relations between events are closely connected with cause and effect. But cause and effect, we said, cannot be shown to exist.

> In order that something like cohesion, something like causality, that some kind of meaning might be revealed and that it can in some way be told, the historian must invent units, a hero, a nation, an idea.
>
> —HERMANN HESSE

The question then remains: on what basis may we judge what happened in the past—and perhaps more importantly, how may we derive from it examples for our present actions or a foundation for our beliefs?

Did Caesar cross the Rubicon? Was Jesus Christ crucified? Was Columbus a hero? Did Marie-Antoinette really tell them to eat cake? What shall we write about colonialism? Sexism? Progress? That is, when we have done our best to appreciate the "facts," how shall we judge their adequacy as history?

We know it from the movies—and from books. We say, "That didn't ring true," or "That was far-fetched"—or, on the other hand, "That was true to life." Similarly, we may judge history on its *truth claims*—namely, whether we think it represents a credible and faithful interpretation of our world or not. In fact, there is no other basis on which we may judge history, unless empirical evidence should incontrovertibly show us otherwise—say, if a grave should exist that is said not to.

At first glance, this might seem to be a negative finding—a defeat in our quest for a valid or objective history. Yet far rather, it suggests positively that every history must be a questioning, a discerning, a reading between the lines, which at its best sets us free from presuppositions. History is intelligent judgment; and truth requires independent, impartial, robust thinking to discern those relations that are sound and those

---

6. Every cause and effect represents a value judgment. $x + y = z$ excludes a, b, and c, and everything else in the world. There is little difference between our choice of such scientific variables and the events we choose to write history. This relates, too, to our earlier discussion of facts.

that are not. This becomes all the more important with the proliferation of fake news and deep fakes.

In fact, if history is *not* a questioning, a discerning, a reading between the lines, then our view of history—personal or collective—may become a hazard to society. If history is thought to be set in stone, it is greatly exalted—beyond what we are able to know. This has led to revolutions and wars.[7]

This all has a bearing on what may be history's most important role of all: its relationship with *religion*.

History and Revelation

There has been, since earliest times, a close relationship between history and revelation. This has not always been an easy relationship, particularly over the past few hundred years.

If we face it square on, we have to admit an element of doubt to every history. This may not matter much when we are dealing with secular history. It may seem adequate that we have a mere balance of evidence in hand or have reached a commonsense judgment as to what happened in the past. Yet this is not a standard high enough for religion. It is inconceivable that religious certainty should be based on any doubt at all or that any act of faith will be, to some extent, a leap in the dark.

There have been three major schools of thought:

- to treat religious history as unassailable fact
- to treat it as fundamentally implausible, or
- to treat it as symbolic—that is, standing for something other than it seems.

There may be a fourth possibility. We may choose to discern what it is in history that we know to be true, apart from the facts so-called—much as we might judge an egg not by the facts that are printed on the carton (the description of its contents or its sell-by date) but by whether it sinks in a bucket or pours in a pan.

There is a proverb that is attributed to the Chinese official Pang Cong, "Three men make a tiger."[8] Which is to say, there is a tiger roaming

---

7. Not to speak of the justification of the *status quo*.
8. Ciyun, *Chinese Idioms*, 121.

in the market, if three men say so. Yet if three men saw a tiger roaming in the market, then there surely was no market. In other words, there is something about the truth that speaks louder than words.

The theologian Reinhold Niebuhr proposed that, rather than proving the certainty of religious history, one might validate revelation (or not) on the basis of its *truth claims*. He wrote, "The truth of faith is correlated with all truths which may be known by scientific and philosophical disciplines."[9] That is, history needs to be credible not merely from the point of view of its factual content but from the point of view of what is credible *in itself*. The entire story must ring true.[10]

We may also put it like this: it must ring true in an open or global system, and this bears different marks of truth than a closed system.

Take the example of a friend who comes to tell me that she was just held up at gunpoint. I assess her story not merely by whether her facts cohere—in fact, she might not seem at all coherent—but by a host of factors emotional, material, social, psychological, which tell me whether she is telling me the truth or not.

Therefore, if religious history is not credible *in itself*, one is not dealing with true religion, no matter what its claims to history may be. Conversely, if religious history is credible in itself, this may give us good reason to believe that we are dealing with true religion.

Another way of putting it is that our *criteria* for judging religious history are critical. The criteria that have been used traditionally—in particular, validating a religious story through supernatural events—have been less than satisfactory. It has been less than satisfactory, too, that we have sought historical certainty. Even if we take "certainty" on balance, it can only be on balance. We need to validate religion by what is credible in itself.[11]

Supernatural aspects of history may, of course, not be credible in themselves, if an anti-supernatural interpretation of history is accepted.

---

9. Niebuhr, *Faith and History*, 152.

10. Francis Schaeffer considered that "from the viewpoint of the Scriptures themselves there is a unity over the whole field of knowledge" (Schaeffer, *God Who Is There*, 117). My argument is a mirror image of this. From the viewpoint of unity over the whole field of knowledge, one may assess the Scriptures themselves.

11. One could call this a theory of explanatory force, yet the explanatory force lies in the knower. Apart from this, one needs to consider whether "credible in itself" means credible in all aspects.

In the coming chapters, we give some further thought to the possibility (or not) of the supernatural.

> **Excursus**
>
> But suppose that we have just read a gripping novel. Would a *novel* not be credible in itself?
>
> In some cases, yes—yet this would depend, too, on the sophistication of the person who writes it and of the person who reads it. It makes a big difference whether writer and reader are naïve or astute, specialists or generalists, and so on.
>
> Further, a novel tends to represent a closed system, while true history draws on a global system. A global system is fed in many different ways and may become credible in a multidimensional way that a novel does not—all the more so as the reader matures.[12]

---

12. The epistemologist W. Jay Wood wrote, "Being justified in a belief is . . . a consequence of my cognitive state's being in right relationship to the world, whether or not I am so aware" (Wood, *Epistemology*, 128).

Chapter 35

Mind-Body Dualism

The mind hovers, so to speak, over the surface of reality.
If this were not so, we would all be mere machines.

WITH THE INSIGHTS GAINED through Bacon in particular, in chapter 32, we now have the goods to explore mind-body dualism. René Descartes first brought this issue to the fore, when he wrote in 1641 that "it is certain that I am entirely and truly distinct from my body."[1] That is, an immaterial mind interacts with a material body.

This separation of mind and body has increasingly been called into question. Above all, a progressively materialistic outlook has led us to wonder whether there is any mind at all. The mind, wrote David Armstrong, is "nothing but the brain."[2] Here, our new view of language may help. But first, we survey the old, commonsensical view of mind and body.

I tap my finger on a tabletop. I drink a glass of milk. I feel the warmth of the sun on my face. Such experiences seem perfectly real to me. So does, say, the passion that I have for my electric pickup, my grief over the passing of my wife, or the fact that I am a New Yorker. Which means that, on the surface of it, my life seems real to me, through and through. Now consider what this means to me philosophically.

It seems, therefore, that I am living in a real world. It is not imagined or illusory. Further, it seems to me that I am an *observer* of this world,

---

1. Descartes, *Discourse on Method*, 115.
2. Armstrong, *Materialist Theory of Mind*, 1.

not merely a robotic presence there. On this basis, it would seem to me that I have a mind that observes reality: mind here, reality there, which separates my mind from the things it observes—and separates my mind from my body, which, too, is a thing.

Understanding the Mind

What, then, is "mind"?

If we separate the mind from the things it observes, it seems difficult to explain how a separate mind should exist in a world where, apparently, only matter exists. Yet if we propose that something *else* exists of which the mind is made, we face the daunting prospect of *proving* it—alternatively, proving that the mind exists in some kind of special *isolation* from the rest of things.

> Physical science has progressed by leaving the mind out of what it tries to explain.
> —THOMAS NAGEL

Today, we do not adequately understand the mind. It presents us with some of the most intractable problems of philosophy and science.[3] Yet we are able to describe the existence of a certain kind of dualism, and it works like this:

In chapter 32, we noted Bacon's observation that definitions consist of words, and words beget words. That is, there are words inside words, and inside those words are words, presumably to infinity[4]—again, not meaning that language contains an infinity of different words but that definitions of definitions of definitions never end. This means that our words and the things that they describe, when we examine them closely, are some kind of *nothings*.[5] Yet this is the antithesis of the something that our reality seems to be.

---

3. Chiefly through our belief in cause and effect, mind-body dualism is a problem. It flags a problem with our view of causality.

4. Since the 1950s, the theory of language has been integral to the study of the mind.

5. The polymath Gottfried W. Leibniz held that matter is "actually sub-divided

Mind-Body Dualism

We live with a strange paradox. Intuitively, we believe in *things* and speak about *things*, yet our very own words and thoughts prove to us that they reach into an endless expanse.

Now this disjoint—namely, our experience of things, on the one hand, and the difficulty of pinning them down, on the other—may explain why we perceive our mind as being so different—also, our many situations and states of mind that give rise to a sense of unreality or detachment, among them: the surprise that there is something rather than nothing, the imposter syndrome, déjà vu, depersonalization, or a sense of alienation.

Ultimately, there may be no disjoint at all. It is only when we seek to describe our existence in terms of different levels of reality—for which we need to create As and Bs—that the problem confronts us and astounds us.[6]

Excursus

How, then, might this contribute to the solution of the mind-body problem?

Perhaps it does not—yet it does integrate the mind-body problem with a philosophy of relations: vanishing relations versus things that seem real. We have (we suppose) a real reality—which is partnered with an unreal language that can never truly get a *grip* on this reality. The very nature of our language curiously distances our words and our mental processes from the reality they describe. The mind hovers over the surface of reality, as it were.

Further, it seems clear to us that, if this disjoint did not exist, we would be mere machines—as would all living beings, we might add.[7]

This has a bearing on our discussion of free will, in chapter 33. If there is no real correspondence between mind and matter—if the mind

---

without end" (Savile, *Leibniz and the Monadology*, 236).

6. Even before one begins to discuss the explanatory gap between physical properties and experience, it is a mistake to isolate them as "things," because there are no As and no Bs.

7. An interesting question is how far down the phylogenetic scale living things are conscious. My own view: a long way down.

"hovers over the surface of reality"—this may loosen the relationship between mind and matter, and strengthen the possibility of free will.

In fact, we may choose to do things we would not do in the natural order of things—say, eat gravel or (try to) fly by our fingers. This in itself points to real freedom of choice.

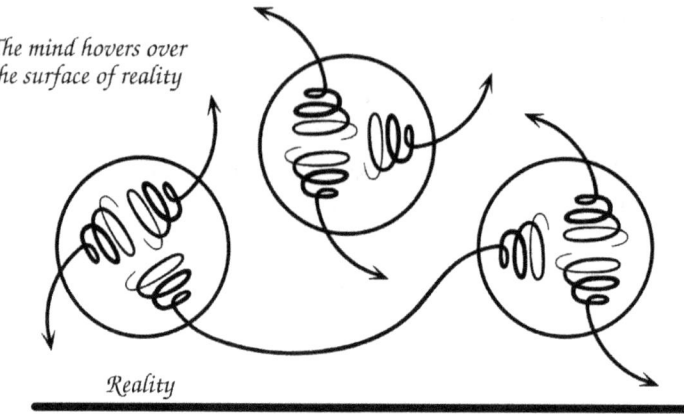

Consciousness and Attention

We have not mentioned it by name, but it has been written all over this chapter so far: mind and body have to do with *consciousness*. It is, however, not so much consciousness which interests us here as *attention*—which is an *apect* of consciousness. In the next chapter, attention becomes particularly important to our analysis of *reason*.

Attention, wrote the philosopher Jesse Prinz, is "conscious awareness."[8] It is the *concentration* of consciousness on some phenomenon, to the exclusion of other stimuli.

When we are *conscious*, we are aware of our surroundings;[9] however, we may not be *attentive* to special aspects of these surroundings.[10]

Consider an imaginary conversation. Pointing to my arm, you ask, "How did you scratch yourself there?" "Oh!" I exclaim. "I don't know. It completely escaped my attention!" Then, with a philosophical turn of mind, you ask, "Were you *conscious* at the time that you scratched

---

8. Mole, "Attention," §21.2.

9. Our surroundings, according to some, include my own body, which surrounds my mind.

10. Simon Blackburn wrote that consciousness is the theater where my thoughts and feelings have their existence (Blackburn, "Consciousness," in *ODP*, 76–77).

yourself?" "Well, of course!" I reply. "At least, *presumably* I was! But not about the scratch."

We become aware of certain aspects of our surroundings sooner than others. We quickly become aware of pain;[11] we are vaguely aware of smell; yet we hardly become aware of what are called visceral interceptors, such as our heartbeat or breathing. We might even ride a bicycle while eating an apple and be fairly unaware of both. We call it the hierarchy of the senses.[12]

We become *attentive*, in any and all of these aspects,[13] when we encounter the *unexpected*—or, we might say, the *contradictory*. Conversely, wherever there are stimuli to which we have become accustomed, we hardly notice them at all.

"But wait a moment," someone might say. "I'm attentive to this book as I read it, but this surely has nothing to do with the *unexpected*."

Yet if the words of this book were wholly *expected*, we would surely not read it. We would say, "That's old news," or "I want something more stimulating."

Picture a pendulum, swinging, swinging, swinging. So little contradiction does this present us with that, rather than producing consciousness in us, we use it to induce hypnosis. But let the pendulum suddenly *drop*, and we jump forward to examine what has happened. The pendulum has contradicted our expectations.

In fact, this happens all the time, in many ways. A shadow passes over my table at a restaurant, I feel a sudden pain under my foot, or there is a strange taste in my coffee. These all contradict what I expect—and immediately, I want to know: "What is it? Why? Where did this come from?"

Ultimately, it all has to do with the relations that we trace between things—whether they contradict the world or coincide with it. Such contradiction becomes important to our discussion of *reason* in the chapter that follows.

---

11. With exceptions like the cut on my arm.
12. The hierarchy of the senses is not, however, universally the same. There are cultural variations.
13. Strictly, "sensory modalities."

Chapter 36

The Purpose of Relations

Our reason, rather than serving as a foundation for thought or representing a constructive enterprise, kicks in to support predetermined purposes.

IN THE WORDS OF the psychologist Kelly G. Wilson, everything depends on the *intent* of our analytic effort.[1] That is, every analytic effort has a purpose for which it exists. When we reason, we reason with a *purpose* in mind. Hence the title of this chapter: *The Purpose of Relations*.

This suggests something we may easily miss. The purpose of reason, at any time and any place, is *given*.[2] Before thought begins, its purpose awaits us. Therefore, our reason is in *bondage* to a preexisting purpose. Let us begin at the beginning.

Intuitively, we assume that our reason is used for *constructing* things. Often enough, we use it (we think) to constuct things *from the gound up*: if not systems of thought—such as schools of philosophy, political orders, psychological theories, and the like—then skyscrapers, computers, space flight, and a million things besides.

We further assume that our reason is fully in *charge*—at least, when we *want* it to be. John Locke was typical of this view when he wrote,

---

1. K. Wilson, "Some Notes," 206.

2. So, in fact, is the data. This renders data—which on the surface of it is purely objective—subjective.

"Reason must be our last judge and guide in everything."[3] In this view, there is no area of my life where reason does not apply.

But such a view is badly mistaken. It leads, above all, to the false belief that reason is a reliable starting point—if not for everything, then in many a sphere.

With this belief, we undertake things that we think are safely undergirded by reason, when, in fact, they are not. But it may be dangerous in the extreme—to embark upon (for example) building, statecraft, educaton, or farming if, unbeknownst to us, we are following a nonrational course.

What, then, is reason?

We build things with it, and we dismantle them with it. We establish facts with it, and we tear them down with it. We use it for good, and we use it for evil. It clears our heads, and it plunges us into confusion. Like an axe in our hands, we use it, yet we seldom contemplate it.

Yet there are a few things we can safely say about reason. We know that we use it to make sense of things. We know that we (puzzlingly) apply it to a variety of seemingly disconnected fields: science, ethics, economics, philosophy—even cooking, hiking, or writing poetry. We know, too, that our reason prevails over our emotions—at least at some level, some of the time. Reason may ignore hunger, reject fear, set aside sleep—among many other things.

Perhaps most importantly, in terms of this philosophy, we know that reason is a *conscious* activity—more exactly, an *attentive* activity. It is deliberate, focused, and aware. This enables us immediately to say a few things about reason:

- Reason is closely related to attention, and attention is a "concentration of consciousness." We concentrate our consciousness wherever we find novelty, discrepancy, and interruption—in a word, the *unexpected*. Reason is, therefore, a reaction to the unexpected.

- We may also put it like this: the unexpected presents a contrast with the background conditions of our existence. In the silence of the night, a dog barks; in the buzz of a shopping mall, a window shatters; in the routine of our daily lives, we receive shocking news. Reason awakes where a contrast is found.

---

3. Locke, *Essay Concerning Human Understanding*, 253. Immanuel Kant held that the legitimate employment of reason is the practical sphere (Scruton, Kant, ch. 5).

- Reason, because it is a reaction to the unexpected and an awakening to contrasts, responds to contradiction. It kicks in where we find a logical incompatibility or incongruity in our situation. This gives rise to the questions why, how, who, what, when.

Reason and Contradiction

Thus reason represents a reaction or response, and whatever it reacts or responds to is prior to our reason.[4]

Consider the example of a house without a roof. We need a roof—yet roofs do not come ready made, *as roofs*. This presents us with a contradiction: the need for a roof, where there is no roof, as such. The best we can do is purchase timbers and tiles. But the timbers are uncut; the tiles are loose. We now find further contradiction: we have the parts that we need, but we need the tools: we need a saw; we need nails. Contradiction after contradiction needs to be solved.[5]

Beginning to think is beginning to be undermined.
—ALBERT CAMUS

We need to find the solution to a complex problem, but our software is inadequate for the task. Contradiction. Our team members need to confer, but we find ourselves in different cities. Contradiction. We need to cross a gorge, but its sides are too steep. Contradiction.

We employ our reason where we stumble upon contradiction.[6] Therefore, reason, being rooted in contradiction, is merely based on the

---

4. This is related to our observations in ch. 4: we are not free to think as we please. We are in bondage to our thoughts.

5. Since the contradiction is between things that exist and things that do not, one may wonder whether this constitutes a contradiction. It agrees, however, with a core definition of a contradiction, "conflict or inconsistency, as between events, qualities, etc." ("Contradiction," in *Collins English Dictionary*).

6. Necessity is the mother of reason.

gaps that we find:[7] the missing software, the need to communicate, the gorge that looms before us. Reason leads us by the nose.[8]

Reason bridges the gaps

For the purpose of bridging or filling the gaps that we find, we employ equations, algorithms, models, frameworks, and so on—and we do drafts, experiments, tests, and various things besides. We *do science*.[9]

Yet while this solves the problems, it does not and cannot *ground* anything.[10] The same is true of everything we do. Whether it be inventing, building, governing, or mothering—even theorizing or moralizing—that fills or bridges a gap.[11] Wherever we do something, the gap is always there. While we may not always be aware of it, we shall find it if we just think on it well enough.

We see this bridging of the gaps, too, on a far larger scale. Casting our mind back over the history of the last few hundred years, we see the early modern period of reason and enlightenment, which saw the great system-builders—Thomas Hobbes, René Descartes, Gottfried Leibniz, and more. It was not merely small or local contradictions these thinkers sought to address. Some sought to address major incompatibilities:

Galileo Galilei reconciled the sublunar and the supra-lunar worlds; James Maxwell united electricity and magnetism; and the greatest of all theoretical physicists, Albert Einstein, melded space and time.

---

7. This is not to say that the end justifies the means. In bridging or filling the gaps, we continue to follow the maxims advanced in chs. 8 and 9.

8. Moreover, our decision may be largely subconscious and could be cultural, so reflecting suppressed inferences or cultural norms. One may ask, too, is reason true, when it serves an end yet does not justify it?

9. While this risks conflating science and the application of it, the point will be understood.

10. The Buddhist monk and philosopher D. T. Suzuki noted that the intellect is thought to be illuminating because it points out ignorance (Suzuki, *Essentials of Zen Buddhism*, 331). Note a parallel with an observation of ch. 26, about what theory is not. If reason is interested in bridging gaps, it will have little interest in what lies outside the immediate challenge.

11. This meshes with ch. 26, "Relating Science," to explain why we select particular theoretical systems to the exclusion of others.

Bernard Bosanquet proposed that reason kicks in wherever we have two competing explanations for the same thing,[12] which is contradiction—while Kant considered that reason is the power of synthesizing into unity the concepts provided by the intellect[13]—which suggests that reason comes into play where such synthesis is lacking. This, too, is contradiction or something of the sort.[14]

Notice, before we move on, that because reason bridges gaps, so does *all* reason, including any system of ethics that is based upon reason—as is the case in this book. It is determined by a physical reality. A housing crisis, say, deforestation, or exploitation—even a conversation with my partner or managing petty cash in a drawer—all begin in the moment.

Yet while ethics is grounded in our physical context and the gaps that we find there,[15] *how* we bridge the gaps is subject to the ethical maxims we developed in earlier chapters. These draw on the widest possible context, namely, the relation of everything to everything.[16]

To put it in a more memorable way, ethics is grounded in rocks and trees, mountains and valleys, "a hard day's night" and dinner plates, and all the *givens* of our existence.[17]

Reason and Holism

Yet not always do we put our reason into gear. Sometimes, it slips *out* of gear, so that we grasp boundless things and relations immediately in all of their totality. We may *suspend* our reason, to see everything with one look. At this point, awe and magic appear.[18]

---

12. Allard, "Bosanquet, Bernard."

13. Wilford, "Review of Ferrarin."

14. This does not imply that reason will always be true. Martin Luther King Jr. observed, "Reason, devoid of the purifying power of faith, can never free itself from distortions and rationalizations" (King, *Strength to Love*, 148).

15. Simone Weil wrote that "at each instant of our life, we are gripped from the outside, as it were, by meanings that we ourselves read in appearances" (Weil, *Late Philosophical Writings*, 22).

16. This would confirm the common wisdom that the greatest virtue lies in who we are, not what we do.

17. It is interesting to contemplate that, if we should plant new civilizations beyond Earth, these may trace relations between things quite differently than what we do on Earth. Beyond Earth, therefore, alien ethics could evolve.

18. This offers a means to unite rationalism with mysticism.

The Purpose of Relations

People the world over have experienced mystical moments through which, in the words of Martin Buber, they have entered with their whole being into the a priori of relation. The theoretical physicist Fritjof Capra described it as being involved in "a gigantic cosmic dance."[19] We find such moments in the grandeur of nature, in a musical performance, in gazing into a lover's face—and, for many, in religious worship. Then, the experience having passed, we reenter the *condition* of things.[20]

Where reason has finished its work and contradiction melts away, there we find that the *holistic* qualities of life emerge that we so greatly value and deeply desire: among them, love, beauty, wisdom, generosity, and purpose—and everything besides, which represents a suspension of reason and particulars.[21]

Yet apply our reason to the holistic qualities of life, and those qualities disappear. Analysis dissolves our every dream.

I am overcome with the desire to fill the world with love—until someone asks me to quantify love. I am filled with awe at a royal pageant—until someone asks me why we go to such needless expense. I feel as though I am in a dream, as I trace a journey into space—until someone asks me why we go up only to come down again. Dreams of greater things can survive only in the absence of cold reason.[22]

---

19. Capra, *Tao of Physics*, 11.

20. The philosopher Plotinus taught that, in ecstatic moments, we may experience reality as a whole (Mommaers, *Riddle of Christian Mystical*, 111).

21. The scientific quest, too, has been described in such terms. "You can recognize truth," wrote Richard Feynman, "by its beauty and simplicity" (Feynman, *Character of Physical Law*, 171). Science writer Kitty Ferguson similarly wrote, "The fragmented picture resolves itself into something of great simplicity, elegance, and beauty" (Ferguson, *Fire in the Equations*, 33).

22. Paradoxically, cold reason, while it undermines more holistic concepts, may also create them. It flags contradictions. Wherever we find contradictions—or rather, wherever contradictions arise for me—there reason kicks in and seeks to solve them. In this way, reason helps us to reduce contradictions in our world and create a world that is one.

Chapter 37

The Teaching of Relations

Education has long been thought of as a holistic enterprise. Yet we need not only diversity but the teaching of *relations* between things.

TEACHING IS SAID TO be about *knowledge*. A teacher, by one definition, "imparts knowledge."[1] But here we meet a contradiction. If "knowledge itself" is deficient, as we said in chapter 29, then surely we make a mistake in teaching it. At least, we make a mistake in *prioritizing* it—to the point of making it focal in definitions of teaching.

The primary reason for the deficiency of knowledge is that it is *exclusionary*—and it is exclusionary because it fails to survey the entire expanse of relations, as a priority. In fact, by and large, it screens out the world. It filters reality through a fine mesh.

A large proportion of our knowledge, we said, excludes unwanted influences on independent variables. x casts away not-x, y casts away not-y, z casts away not-z. For this reason, the impartation of knowledge does *damage* through the things it leaves out—damage to ourselves, our society, and our environment.

Paradoxically, the *interrelatedness* of our world has rapidly grown in importance, while the subject that has traditionally *studied* it—which is metaphysics—has largely been sidelined. The "divinity or school

---

1. Some definitions may include competence or virtue. Yet in reality, competence is little taught, and virtue even less.

metaphysics," about which David Hume wrote,[2] is virtually unknown in schools and universities today.

However, in our (post) modern world, expansive thinking is imperative. Pope Francis noted that a fragmentation of knowledge has often led to a loss of appreciation for the whole, a whole that he called "the relationships between things, and the broader horizon."[3] While such fragmentation of knowledge has been helpful for concrete applications, it makes it hard to find ways of solving the more complex problems of the world. It is crucial that we develop a more integrated or holistic thinking.

Education, in name at least, is about holism. In ancient times, there was a saying, best known in Latin: *mens sana in corpore sano*—a healthy body is a healthy mind. It is attributed in different form to Thales, around 600 BC,[4] and reflects a long-held belief that education is both about a healthy mind and a healthy body. That is, education is, in some sense, a holistic enterprise, which aims to produce well-rounded citizens.[5]

For this reason, school courses will typically include both academic subjects and sports—as well as contextual experience, such as educational excursions and contextual learning. Less obviously, one's education will include what the educationist Alan Paton described as "the great silent forces of example, and of laws and customs unconsciously adopted and obeyed."[6]

An early philosopher in the field of education was Plato. For Plato, education was ultimately about social and individual *balance*.[7] It was Plato's outlook that later shaped the education of the medieval universities of Europe. These universities taught the seven liberal arts: the *trivium* (grammar, rhetoric, and logic) and the *quadrivium* (arithmetic, music, geometry, and astronomy). Advanced students often went on to study philosophy and theology.

---

2. Hume, *Enquiry*, 107.

3. Francis, "*Laudato Si'*," §110.

4. Bailey, "Healthy Body."

5. Timothy Leary called it "the old dream of a *universitas*, a synthesis of human knowledge" (Leary, *Politics of Ecstasy*, 189). His reference to knowledge, however, suggests that his *universitas* was more limited in scope than what is intended here. In Bloom's taxonomy, education has three objectives: the cognitive, affective, and psychomotor (Bloom, *Taxonomy of Educational Objectives*, 7).

6. Paton, *Diepkloof*, 98.

7. Plato likened education to "the proper tuning of the instrument," as a precondition for citizenship and philosophy (Plato, *Republic*, 364).

Even today, basic features of the ancient Greek and medieval systems remain. Most simply, our schooling is still divided into separate compartments, with our classes about the same in number. The content of our education, too, still overlaps with medieval subjects—the three R's being examples: "reading, 'riting, and 'rithmetic."[8]

Due to the size and momentum of this educational model, it seems unlikely that we shall see sweeping changes in the foreseeable future. However, there are compelling reasons to rethink it.

What We Left Out

In previous chapters, we gave some pointers as to how we may pursue a greater holism. We said that we discover, on the one hand, relations in the world around us; on the other hand, worlds of relations in other people's minds.

In the case of the relations in the world around us, our minds should range through all the world, to relate things broadly and truly. In the case of the relations in *other* people's minds, we should understand people's inner worlds—for which we need a receptiveness or sensitivity to people's semiotic codes: their utterances, gestures, postures, and so on.

With this in mind, our educational systems place a great emphasis on the world *around* us yet are weak in the area of human understanding and rapport.

In the area of *fact*, most of us are well grounded—whether this be in physics, mathematics, biology, or geography, to give but a few examples. Yet in the area of our daily human existence and experience—personal, social, and global—school curricula are largely empty, and ethics is often taught only implicitly. In terms of a philosophy of relations, our education needs to place a greater emphasis—an *explicit* emphasis—on human understanding and rapport.[9]

Education may further fall short of a rounded or holistic enterprise through the traditional *compartmentalization* of subjects—that is, their relative isolation from one another. The philosopher and sociologist Edgar Morin, renowned for his work on complexity, wrote, "Our compartmentalized, piecemeal, disjointed learning is deeply, drastically

---

8. One speaks of a *core curriculum*, which may, however, vary.

9. John Locke wrote that the most important aims of education are "Virtue, Wisdom, Breeding, and Learning" (Locke, *Educational Writings*, 105).

inadequate." It is one thing to offer an assortment of subjects to learners, quite another to relate these subjects to one another.

> The value of an education [lies in] the training of the mind to think something that cannot be learned from textbooks.
> —ALBERT EINSTEIN

Often, we fail to consider how one subject has a bearing on the next—even less, how it is related to the wider world. While there is a diversity of subjects on offer, the end result is not a healthy holism.

We may recognize, for instance, that a Spanish class has something to do with an English class yet fail to understand how both might have to do with the transmission of culture. We may see that our mathematics class has something to do with our physics class yet fail to understand how both are related to unwanted externalities. We may recognize that our history class has something to do with our religion class yet fail to understand how both are related to knowability.

It is only as we deal with subjects in their widest context that we are able to develop a truly balanced education. More than this, we now have empirical evidence that an interdisciplinary education is related to creativity and genius.

An interrelated education may further address a central concern of alternative educational models, models that often highlight the need for more problem-solving and critical thinking. In fact, by relating different subjects to one another—and ultimately, all things to all things[10]—we are engaging in problem-solving rather than mere learning.[11] We are teaching *relations* more than we are teaching facts. Perhaps rather, we are teaching relations with a wider reach. Pedro Domingos wrote, "In relational learning, every man is a piece of the continent, a part of the main."[12]

---

10. An editor challenged me: "How do we relate a Spanish class to a math class?" The professor of didactic and mathematics Luis Rico wrote, "There are ways of thinking and doing math, based on [Spanish]" (Rico, "Spanish Heritage," 333). Even Spanish and math are related.

11. The theoretical physicist David Bohm considered, "The ability to perceive or think differently is more important than the knowledge gained" (Horgan, "David Bohm," para. 14).

12. Domingos, *Master Algorithm*, 228.

Thinking on our earlier discussion of reason, we noted that reason builds everything backwards, so to speak. This is problem-solving, rather than the artificial construction of subjects in our minds.[13] The nature of reason has everything to do with the nature of education.

Not least, a good education has much to do with the *limits* of our knowledge—and power. We need to teach our limits, too. Knowledge, wrote Confucius, is to know both what one knows and what one does not know.[14] We may add: both what one can do and what one cannot. If we suffer any confusion here, we set ourselves up for disaster.

New Bearings

For the majority of children today, education is compulsory, where once it was the preserve of the few. By and large, therefore, children experience a common introduction to education. The need to adapt to individual students would therefore seem to be smaller than it was.

At the same time, many proposals for the *revision* of education focus on individualized attention—among them indigenous education, progressive education, and various models of home schooling. Which then should it be? A standardized education or a personalized one?

In our discussion of person ethics, we noted the importance of tracing relations from that point in time and space where we find ourselves now—which is the only place to begin. Similarly, we can do no better in education than to acknowledge our subjective reality and to start from there. Therefore, education should always be personalized—as far as possible beginning with the knowledge and experience of each pupil.

Finally, it is a truism that education may be used for ideological or propagandistic purposes. History is peppered with examples of ideologically charged education—*radicalized* education—and often enough, the results have been tragic.

In terms of a philosophy of relations, we have described ideology and propaganda as limiting the full scope of the *relations* that are available to us—with the result that the relations that we trace between things

---

13. Plato considered that education is not teaching but a recollection of what the student already knows (Mayo-Wilson, "Plato's Theory of Recollection"). John Dewey proposed that human thinking is essentially a matter of problem-solving (Denton, "Problem-Solving," 382).

14. Socrates said that he was wiser than those reputed to be wise, because he did not claim to know what he did not know (Bett and Morrison, "Socratic Ignorance").

may encompass only narrow or partial points of view. Conversely, the opening up of the entire scope of relations withers ideology and propaganda. Therefore, rather than there being subjects that are off-limits in education—politics, ethics, or religion, for example—it is our *approach* to these subjects that holds the danger or the blessing.

An understanding of ideological subjects is critical to our education as citizens, and any curbs on them may be more dangerous than the teaching of them, in an age where political sophistication and new ethical bearings are urgently required.

Part VI

The Ultimate

We define the ultimate as the final principles. The sixth part of this book addresses *transcendent* themes, which lie beyond or above the realm of our normal human experience—yet we seek not to exceed what can reasonably be said.

Chapter 38

Religion as Relations

How may we reconcile religion with reason? We here consider the truth claims of religion, based on the way that religion traces relations between things.

PHILOSOPHY'S INFLUENCE ON RELIGIOUS thought cannot be underestimated. More recently, one thinks of Ludwig Wittgenstein, who had *and has* a profound influence on post-liberal or narrative theology. One may add such names as Karl Marx, Alfred North Whitehead, Michael Polanyi, and Jacques Derrida—in fact, a great many more—who have deeply influenced religious thought.

At the same time, religion has had a major influence on philosophy. In fact, philosophy may (arguably) be described as a reaction to theology. Terms such as atheism, secularism, dualism, (non) rationality, and many more are closely related to theology.

At the same time, it is important that we do not discuss philosophical ideas in theological terms or conflate philosophy with theology. If we use religious ideas to judge religion, then religion becomes its own judge.[1]

Religion raises any number of questions. Why is it that people, in great numbers, worship abstracta? Why is it that they believe in immaterial causes? Why do they permit the erasure of reason? Why do they

---

1. Even if religion is said to be based on revelation, it is always influenced by human thought, which deserves contemplation.

entertain distinctive grand narratives: for example, sin and redemption, or struggle and reward?

Today, by far the majority of our planet's population claims to be religious. Whatever we may think of religion, there has to be something in it. We dare not assume that there is no valid thinking there—even in the case of supernatural claims that are widely believed. If we truly want to take account of everything, briefly, we dare not pass over religious beliefs.

Truth Claims

We begin here with the bare existence of religion and how this relates to society and a system of government—above all, secularism, which is the principle of separation of the state from religious institutions.

Every major religion has a holy book. Whether it be a solemn black Bible, a Quran adorned with gold leaf, or a colorfully illustrated Bhagavad Gita, few people today will not have opened such a book and looked inside. In fact, religion, more often than not, is *synonymous* with a holy book.

In terms of this metaphysics, religion's holy books—whether we accept them as true or not—represent the ordering of relations in this world—and perhaps, too, out of this world, as we shall consider shortly. Religion, therefore, comes down to something far simpler than we ordinarily imagine. As in the case of various fields of inquiry, it comes down to *things and relations*. Religion influences the way that we trace relations in this world.[2]

By so tracing relations in this world, we hold up the world in our minds to the world itself. This takes us back to chapter 10, where we traced our actions back to *mental models*. Religion's holy books create mental models, and mental models motivate people. Therefore, religion's holy books are not mere curiosities. They have the power and potential to bless or to destroy.

Of particular interest in this chapter is how this affects the relationship between religion and state—in a word, the issue of *sovereignty*. If we do not address the issue of sovereignty, we leave the door open to *religious* sovereignty—the rule of Christendom, say, or Islam, Hinduism, or any of the religions of our time. This could invalidate much that we

---

2. Conversely, the way that we trace relations in this world influences our religious inclinations. Most importantly, reductionistic thinking tends to exclude God—something to which we return in ch. 39.

have said about public ethics so far—above all, the need for a *syntocracy*, which means a participative process, the priority of relations, the protection of identity, and so on.

The relationship of religion and state is, of course, not merely academic but has been vigorously fought over—and is still fought over—the world over.[3] Centuries of religious conflict in Europe in particular led to the realization that beliefs and practices that are seen to be justified by God may have fearful consequences. Out of this arose the time-honored idea that the dangers of religion may be attenuated by reason. Where differences of belief seemed irreconcilable, it was thought that reason might command universal assent.

This led to a practical solution, which is widespread today, namely, the separation of church and state—now state and religion—or, to put it another way, secular states that grant religious tolerance.

With this in mind, we here consider the *truth claims* of religion—which are propositions central to religion, which we can everywhere, at all times, judge to be true. If we can establish such truth claims, we may have a basis for the harmonious coexistence of religion and state, where both are agreed as to truth.

From our earlier ethical maxims, we may derive three truth claims. By this, we mean core beliefs with philosophical foundations, which touch on core issues of religion. The words that we use to describe these truth claims are not easily selected and require some qualification:

Harm. We have said that we may trace relations in this world that are expansive and integrative, balanced and broad—or short-sighted, self-interested, and parochial. The latter lead to behaviors that are destructive or only *seem* to be attractive, so that we enter crises of dead ends—so damaging our environment, society, and selves.

We shall call this "harm"—which refers to any injury, suffering, distress, or destruction we bring upon the world.

We may note again that harmful tendencies may be *inherited*. In chapter 10, we said that the very words that we speak may represent the transmission of conceptual relations through many centuries. Our cultural inheritance, which, in many ways, represents the wisdom of the ages and enables us to survive and to thrive, will, in other ways, drive us to

---

3. What is arguably new is that it is not merely fought over within states, but the struggle is exported—most visibly through terrorism but in many other ways.

ruin: through war, the destruction of the environment, or the subversion of familial relations, among other things.

Harm, therefore, suggests something that theologians call *original sin*—which is to say, that a corrupt human nature is inherited—or, as the theologians have it, inherited from the start.[4] We *inherit* ways of thinking and doing that do much damage in our own day.[5]

Ownership. Not only do we trace relations in the world *around* us. We trace relations that help us find a reasonable estimate of our own place in the world. In chapter 8, we presented the maxim *Know how you yourself are situated in the world.*

Without a true estimate of our own place in the world, it may be hard to bring into focus even the basic character of our existence. Has my view of the world been balanced and broad, or has it been short-sighted, self-interested, and parochial? Has it harmed me, or have I harmed others, my society, or my environment?

Too often, the relations that I myself trace in this world—that is, the way that I arrange the world in my mind—has *harmful* effects. Therefore, I need to come to terms with who I really am. I need to accept it, admit it, and own it. I need to set aside pretence and illusion.

We shall call this "ownership"—which refers, above all, to the ownership of the harm that we do.

Here we may note that "ownership" applies both individually and collectively. In many cases, we are *unable* to harm others individually. Institutions, traditions, systems, even mobs may be necessary to inflict certain kinds of harm.

Abandonment. Combine this, thirdly, with an earlier observation, that in many cases, I cannot rearrange the relations in my mind—that is, modify my worldview—unless I am willing to renounce every relation I have ever traced. If I should value anything more highly, this will tend to restrict, confine, reduce, and distort the truth. To put it another way, my

---

4. One alludes to original sin. However, we are not in a position, philosophically, to claim that a corrupt human nature was inherited *from the start*—only that it was inherited.

5. This agrees, too, with a principle that lies at the core of much of this metaphysics: the relations that we find in this world are almost infinite, yet we ourselves are finite. Therefore, our tracing of relations is always incomplete and partial, and all models are wrong.

view of what is true may be distorted by an unwillingness to abandon the way that I currently relate things one to the other.

The first step of philosophy, we said, is *abandonment*, and such abandonment must challenge everything I have ever known. That is, it is only when the totality of the relations that I trace is brought into flux that I myself am truly changed.

We may think of a world-class interior decorator, whom I engage to remake my sitting room overlooking New York's Central Park. "But that painting," I say, "must stay." The interior decorator is nonplussed: "It's a dark, old painting," she says. "I can't implement my vision with *that* in the room. It's all or nothing."[6]

Any religion, to be broadly compatible with sound philosophy, will not be about piecemeal changes to my personal life or artificial changes, externally induced. To borrow a term from Hegel, it will not be a "positive" religion: an objective system of rules and laws, dictated by people or gods.[7] In short, it will not be about legalism, not to speak of law. It will bring about deep changes to the way that I conceptually arrange my world—and, with that, will produce deep motivational and attitudinal changes.

We shall call this "abandonment," where this is the abandonment, even repudiation or rejection of that which I held before, and a deep and true turning.[8]

Religion /rɪˈlɪdʒ(ə)n/ *noun* The belief in and worship of a superhuman controlling power, especially a personal God or gods.[9]

---

6. The Native American proverb implies all or nothing: "When you discover that you are riding a dead horse, the best strategy is to dismount" ("Tribal Wisdom," para. 1).

7. Hegel, "Whence Came the Positive Element [in Christianity]?," in *Positivity of Christian Religion*, §4.

8. This has the necessary corollary that there is ultimately only one way in which we may do harm in this world, and that is through our failure to reach a place of abandonment. Anything else we call harm—our individual offenses, subversions, or transgressions—is an illusion. It is only the absence of abandonment that prevents us from doing harm.

9. "Religion," in *Oxford Dictionary of English*.

We now have three religious truth claims that we derive from a relational philosophy: harm, ownership, and abandonment. In fact, all three of these may be subsumed under the concept of "awakening," which we called a disintegration of our first philosophies in chapter 23.

If this analysis is correct, then religion is not primarily an experience, a set of values, a cultural phenomenon, a manifestation of the unconscious, a collection of metaphors, or a language game—among other things. Rather, one may ask whether it bears three simple marks:

- Does it recognize the harm that one does?
- Does it set aside pretences, through ownership of such harm?
- Does it require abandonment?—namely, a deconstructing and reordering of every relation one has ever traced?

But we have made a vital omission. A religious person would want to know how a god *responds* to such attitudes and actions on our part. Do the attitudes and actions lead to salvation? Do they bring sanctification? Do they find God's blessing?

While it would be tempting to try to answer such questions here, this may require the information of special revelation, which traditionally lies beyond the scope of philosophy.[10] We shall, however, touch on certain *aspects* of these questions in chapters to come.

The Secular State

A state commonly grants its citizens the right to freedom of religion, freedom to *change* their religion, and freedom to *manifest* their religion, in teaching, practice, worship, and observance. Yet religion, in this view, should not transgress the authority of the state.

This we find to be in agreement with a philosophy of relations. Religion is not an objective system of rules and laws. Rather, it rests on the arrangement and rearrangement of the world in one's mind, to which we referred as abandonment.

Thomas Jefferson, an early president of the United States, put it like this: "The legitimate powers of government reach actions only, and not

---

10. The extent to which revelation lies beyond philosophy was discussed by theologians Augustine, Thomas Aquinas, and Bonaventure, among many others.

opinions."[11] Therefore, while there is freedom of religion, all citizens must yield to secular law, as it affects their actions.

However, there is a *danger* in Jefferson's view. If we understand religion as we do any other worldview—namely, as a picture that we hold up to the world, which directs our actions in this world—then religion is not about "actions only." Such a view would not reflect a true state of affairs and would not, in itself, render religion safe—or safe for the future.

Rather, religion needs to be assessed in terms of its *truth claims*. Without a contract between citizens and state, through which citizens submit even religious truth claims to reason, one risks powerful incompatibilities.

At the same time, we need to separate truth claims from the *supernatural* claims of religion—or the supernatural *superstructure*, as some have called it. If we should decide against religion's supernatural assumptions—in the sense of rejecting them in principle—then the danger of religious persecution looms.

The very notion of truth claims reverses the popular conception of religion as a supernatural superstructure that validates its truth claims. Rather, truth claims validate (if they do) a supernatural substructure. Blaise Pascal wrote, "Miracles are the test of doctrine, and doctrine is the test of miracles."[12]

Most basically, if a religion's truth claims potentially usurp the powers of the state or contradict an open society, then religion must either yield to the state or be treated as fundamentally seditious.[13]

---

11. Jefferson, *Writings*, 400.
12. Pascal, *Thoughts*, 261.
13. One needs to be thoughtful about what this really means. In the past, there have been major struggles between religion and state: over conscientious objection, tax exemption, women in society, marriage law, home schooling, religious symbols, and more. Theologians make a helpful distinction between submission to the state and subjection to the state. Submission may include respectful noncompliance, which can lead to voluntary suffering on the citizen's part; while subjection is unconditional and must be accepted or rejected. Respectful noncompliance should be treated with compassion.

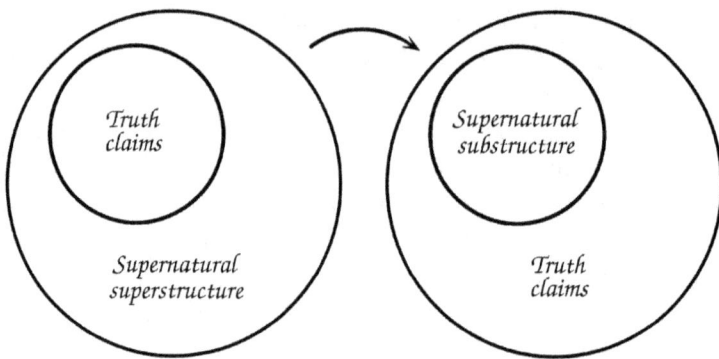

We may view it all another way. The relations that religions trace are typically bounded and limited. While it is true that religion generally makes a claim to every area of life, because everything falls under the power and authority of God, this does not provide the means to trace relations between all things in the broadest possible way.

Where religious authorities take significant control of a nation or region, they govern from the limited perspectives of religion. Further, they may too easily claim divine right, and divine right might substitute for thought. With such a reduction of relations, we are all imperiled.

Chapter 39

God beyond Relations

> A God who is causally related to the universe will be pushed back. This need not mean that God does not exist.

THE DEBATE ABOUT THE existence of God has been vast. There have been arguments ontological, cosmological, teleological, empirical, inductive, deductive, and subjective.

Ultimately, however, the existence of God must come down to the simple question: is God (we shall use the familiar "he")[1] causally involved in the universe?[2] Everything that we say about the existence of God or his revelation—if there should be such a thing—may be reduced to this.

As our understanding of causality has grown, so God has been increasingly removed from our understanding of the world. "It's a shrinking kingdom," said the actor and writer Stephen Fry.[3]

There have been some symptomatic markers along the way.

- Soon after Isaac Newton formulated the laws of motion, Paul-Henri d'Holbach scandalously (at the time) applied Newton's principles

---

1. In the words of Rabbi Aryeh Kaplan, "We refer to G-d using masculine terms simply for convenience's sake" (Kaplan, *Aryeh Kaplan Reader*, 144).

2. A noninterventionist God is not entertained here (such as the deist's God), for reasons that will become apparent in a moment.

3. Fry, in Peterson, "Atheist in the Realm," 1:28:39.

to *humanity*. He wrote, "[One] is unceasingly modified by causes."[4] With this, God was exiled from the world of personal providence.

- A century later, the emperor Napoleon Bonaparte reportedly complained to scientist Pierre-Simon Laplace that he had not mentioned the creator in his system of the universe. Laplace replied, "No, Sire, I had no need of that hypothesis."[5] With this, God was banished from the system of the universe.

- A century later, the naturalist Charles Darwin wrote in his autobiography, "The old argument of design in nature . . . fails."[6] With this, God was exiled from the realm of nature.

- Finally, the theoretical physicist Stephen Hawking wrote, "So long as the universe had a beginning, we could suppose it had a creator."[7] The evolutionary biologist Richard Dawkins drove this to its logical conclusion. God was not left with much to *do* then: "Just set off the big bang."[8] With this, God was dismissed from any ongoing involvement in the universe.

Faith in God, in order to exist, must set aside causes. It is only a life that exists beyond the world of natural causes, as we know them, that holds room for God.[9]

Every physical cause we discover or invent drives God back and banishes him from some aspect of (some call it) his creation. God is thus removed from the *explanation* for things.[10] Today, we can no longer explain the creation as a manifestation of God's power or glory. No longer can we explain a confluence of events as God's gracious providence or intervention—alternatively, his judgment. We now explain it all in terms of natural causes.

Yet even as we have magnified natural causes, our thoughts about them have undergone a steady transformation. Beginning in chapter 2, we described a growing awareness that everything is *interconnected*.

---

4. D'Holbach, *System of Nature*, 1:134.
5. Hahn, *Pierre Simon Laplace*, 172.
6. Darwin, *Autobiography*, 87.
7. Hawking, *Brief History of Time*, ch. 8.
8. Dawkins, "Improbability of God," 9.
9. The familiar argument that God is the first cause, or the cause behind all causes (the cosmological argument), is invalidated, therefore, at least in its current form.
10. Much the same is true if we have any causal influence on God.

With this, causes and effects have become boundless. They are beyond knowing. Our powers are vastly inadequate in tracing all the relations around us. We saw this in earlier chapters—among other things, through examples of the mulberry tree and the pond.

In important ways, this returns us to a premodern outlook: a seamless integration of earth and sky and sea—a vast interrelatedness, which can no longer be held by the laws of nature and their outcomes.[11]

> The simple adoration of a God has preceded
> all the systems of the world.
>
> —VOLTAIRE

Not only this. We noted earlier our inability *altogether* to prove that causality exists.[12] There are, as we put it, no As and no Bs: no things, no objects, no entities, no events, and so on. Rather, they do not exist *in themselves*. They have no *hard core*. They do not cohere. They all dissipate in the endless expanse. In the final analysis, we cannot speak of causes.

Now if we cannot speak of causes, then we no longer drive God back—either from the world around us or from our personal lives. We no longer bring *explanations* to our world.[13] Cause and effect dissolves, and we enter with our whole being into "the *a priori* of relation."[14] A holistic thinking emerges, which may be impossible to distinguish from faith in the omnipresence of God—in fact, the personal *providence* of God.[15]

God? How can we speak of God, where we have just done away with things, objects, entities, events? Above all, causality, which would be central to his existence?

---

11. Cosmologist and theoretical physicist John Barrow noted, "It is one thing to know the laws of Nature, but quite another to know the outcomes of those laws" (Barrow, *New Theories of Everything*, 66).

12. It seems extraordinary that this possibility opens up for us. It might not, after all.

13. One might say that we are now surrounded not by particulars but the universal.

14. Buber, *I and Thou*, 27.

15. Martin Buber intimated much the same: "So long as the heaven of Thou is spread out over me the winds of causality cower at my heels, and the whirlpool of fate stays its course" (Buber, *I and Thou*, 15).

Suppose that I have written a book. After three long years, I wrap it all up and put out proposals to publishers. In three days, a commissioning editor contacts me. He will take my proposal forward for a decision of the publisher.

I am excited. I couldn't have *dreamed* of this. It came about, I tell myself, because I enrolled in some very valuable courses—and then there was the author who examined my manuscript—and my partner, who encouraged me to cast doubts aside. But wait, I cannot speak of causes. I am defining As and Bs.

Of course, there is a long way to go yet to publication. What if the publisher doesn't like the managing editor's pitch? What if they finally get cold feet? What if my book should fail to sell? What if . . . but wait, again I speak of *causes*—of As and Bs.

The approval of the commissioning editor, and all that will result from it, is caused by *nothing*. No, it is caused by *everything*. It is caused by everything and nothing! Yet it cannot be nothing. This is the convergence of everything! For a moment, I feel the glow of mystery and harmony—and unity.

Excursus

Yet what does it mean, "it is possible" that God exists? *How* possible? Is he *remotely* possible? Is he all but certain?

It means that a belief in God would be completely compatible with a philosophical view of the world—just as, say, the existence of birds would be completely compatible with the idea of an atmosphere—whether or not one sees any birds.

We may note, however, that there is a great difference between possibility and necessity. At this stage, we have not established any reason why God's being there should be *necessary*. Unless it is necessary, it all remains fairly academic.

Paradoxically, the loss of God as an original cause—and our return, in a sense, to a premodern outlook—may make the possibility of God's existence stronger. We can no longer explain God away.[16] Such

---

16. Can we then explain poltergeists away? Or demons, fairies, or any number of

escape from causality renders it *possible* that God exists and *possible* that religious conceptions about him are true.

It seems that a sense of God's providence could not differ from this sense of all-encompassing mystery and harmony.

This raises a simple question. If we say that, ultimately, causes are not real, how, then, should we entertain the personal *providence* of God, which would require that God does have a causal effect in this world? Does God not *cause* providence to happen?[17]

Clearly, God would have to be involved in this world beyond the kind of causality we know or *imagine* we know—that is, beyond the relations between things that we typically trace.

One may use the analogy of a chess game. A careful observer reports that the entire game is governed by the strictest of rules. Yet there is clearly, at the same time, something happening at a completely different level, which is the thoughts and actions of the players. Whether their thoughts and actions are free or determined—or whether the players exist at all—one will never know merely by examining the rules.[18]

Even so, without seeing the players, one could assume that they must exist—at least, something *like* them—because a chess game plays out differently, and intelligently, every time. There is clearly something there that is more than the rules.[19]

In fact, we see something like this in our world. In just the same way, we are at a loss to explain the complexity of our universe, where the rules should produce either complete regularity or complete chaos.[20] The theoretical biologist Stuart Kauffman wrote, "We have no adequate theory for why our universe is complex,"[21] while Stephen Hawking wrote, "Why is it

---

mythical creatures? It depends whether they are necessary to our existence. This question applies both to them and to God, and is addressed shortly.

17. This includes the question as to whether God speaks in this world.

18. There is a parallel with the analogy of ch. 32, about the clay pigeon and the church bell—causes and effects that are chosen both by rules and by choice.

19. Kapila entertained such a view: "It may be that its fine nature makes it imperceptible" (Ambedkar, "Kapila," in *Buddha and His Dhamma*, §2).

20. This recalls the argument from design—namely, that there has to be an intellect behind the universe. However, this does not ask, "How much design?" Judging by the statements of Kauffman and Hawking, the answer could be: almost none.

21. S. Kauffman, "Why Is Universe Complex?," para. 2.

not in a state of complete disorder at all times? After all, this might seem more probable."[22] There seems to be more to it than the rules.[23]

The idea of a different level of reality has been seriously entertained by various scientists. Science writer Natalie Wolchover summed it up, "Space-time may be a translation of some other description of reality"[24]—and while she was not proposing the existence of the supernatural, the idea could conceivably open the *way* to the supernatural.

This may have a bearing in another way on the existence (or not) of God. It provides a cosmological argument—yet not as we have thought of it in the past.

We have just said that the existence of God is not proved but disproved, if cause and effect are omnipotent. Combine this with our observation in chapter 33, that causality, as we know it, does not exist. What, then, causes anything at all in our world to move? It would have to be a cause or causes *beyond* causes.[25] Or rather, it would be a cause or causes that *transcend* all causes.[26]

This is not to say that we believe George Berkeley, who believed that only God moves anything—but it would be something beyond what we ordinarily think.

God's Providence

We asked in chapter 10 what it is that *motivates* me—say, to plant a garden, to embark on a career, or to go to war. Today, there are few who disagree that, most basically, I am motivated when I hold up the world in my mind to the world itself—and find a difference there.

---

22. Hawking, *Brief History of Time*, 156.

23. This is to put it cautiously and conservatively. Everything we see would seem to have more to it than the rules.

24. Wolchover, "Different Kind of Theory."

25. If this is true, there are behavioral consequences. Any view that reckons with causes only as we know them, explicit or implicit, is false. For example, "This might end in failure" (which is causal thinking), "This would be to my advantage" (causal again), or "What will they think of me?" (causal). Since such thoughts are based on a nonexistent causality, we ought not to entertain them. In short, we should act on principle. This includes the maxims of chs. 8 and 9.

26. One is at risk of proposing two levels of causality. But it may all be one. That is, two levels of causality contained in one. Aristotle's fifth cause may suggest this ("Aristotle," para. 2).

God beyond Relations

But we may state the opposite, too. Just as a disjoint between expectation and reality motivates me, so a *lack* of such a disjoint *demotivates* me—and may potentially remove all motivation altogether.

A low-level idea now leads us to a fundamental question of our existence, which is "the problem of action."[27] Why should we carry on? If there is no difference between the world in our mind and the world itself, we are robbed of motivation.

Many start out in life with high hopes, pleasant dreams, a positive view of humanity, and enthusiasm to spare. But as we progress through life, the world, in one way or another, breaks us. It breaks us not so much through the hardships it brings to bear on the body—if that should matter at all—but because it assails the mind and emotions.[28]

Disillusionment sets in—disillusionment that presumably means coming to see things for what they *are*.[29]

As we grow and mature, we may come to see the world as a place where hopes wither, dreams die, and energies are sapped. We may become jaded, tired, and disinterested.[30] Søren Kierkegaard wrote, "Ultimately, one becomes tired of everything."[31]

Paul Tillich recognized the problem: "If life is as meaningless as death, if guilt is as questionable as perfection, if being is no more meaningful than non-being, on what can one base the courage to be?"[32] The problem occurs, wrote Tillich, when one *equates* life and death, guilt and perfection, being and non-being.

---

27. Teilhard de Chardin, *Phenomenon of Man*, 226.

28. The religious scholars David Lawrence Edwards and David Lloyd Edwards wrote that a core problem with atheism is "the need which many people feel for faith and a supernatural consolation in order to endure" (Edwards and Edwards, *What Anglicans Believe*, page number unavailable).

29. If brokenness were something piecemeal in one's life—if we were disillusioned only in bits and pieces—the rest might still sustain us. Yet disillusionment settles upon us as a cloak that covers everything. Our mind sees the implications of one disillusionment for all others and grasps the consequences.

30. Philosophy professor Roe Fremstedal observed that there is a hope that occurs spontaneously in youth, which is often disappointed in time (Blöser and Stahl, "Hope," §2.5).

31. Kierkegaard, *Either/Or*, 286. The novelist Ernest Hemingway wrote that the world "kills everyone"—in particular, "the very good and the very gentle and the very brave" (Hemingway, *Farewell to Arms*, 216). Athlete and activist Christopher Bergland said that "motivation fizzles with age" (Bergland, *Athlete's Way*, 198).

32. Tillich, *Courage to Be*, 175.

And so the *disjoint* is lost, and we lose our motivation.[33] With no world to *hold up* to the world, because we have finally seen the world for what it is, motivation dies—ultimately, *all* motivation dies—because motivation is the unexpected, and little is left of it where disillusionment sets in.

What, then, to do, when we are broken? How may we restore any motivation at all, when we have come to see the world for what it is? In fact, how can the world itself continue, where there is no reason for it to do so? It *ought* not to go on—and yet it does.[34]

Excursus

Someone might object, "This is *psychology*, not philosophy."

Not necessarily so. It rests on what we previously observed about motivation. Motivation, we said, springs from the arrangement of the world in our mind, when we hold this up to the world itself. Where we then discover a disjoint, we are moved to act.[35]

Now we have added that, if this disjoint is erased, this is ruinous to our motivation. And because this has to do with the arrangement of the world in our minds, it is as much a rational argument as an emotional, psychological, or spiritual one.

For many, the answer to human disillusionment has to lie *beyond* this world—and though this may be an appeal to consequences—namely, the argument that it *must* be so—indeed, it must.[36] The philosopher Albert Camus wrote, "Where there is no hope, it is incumbent on us to

---

33. There is a concept like it, in rhetoric and philosophy, although its meaning is not as strong: we reach an aporia—a strait—where no action or word will help us.

34. One may refer to the lyrics of "The End of the World" (Dee, "End of the World," verse 1).

35. One may again question the universality of this principle. How could a disjoint lead to, say, my reading this book? In such cases, it is helpful to ask, "What if there had been no disjoint?" No disjoint between this book and one that had been read before, no disjoint between the reader and their world, and so on.

36. This, therefore, explores an empirical proof for God. By way of contrast, "Most definitions of 'God' are such as to preclude any explanation in terms of what can be directly experienced" (Thompson, *Teach Yourself Philosophy*, 62).

invent it."[37] We cannot go on with a view of the world that is born of the world itself.[38]

Religious writings reveal at least two possible ways in which the answer may lie beyond this world.

Firstly, one may strive not for things that are this-worldly but for things that are other-worldly. In various traditions, it is central to religious thinking that "if we live, we live for the Lord, or if we die, we die for the Lord."[39] We continue to strive, yet we strive for something that is not of this world.

What does it mean, then, to strive for something that is not of this world?

Among other things, this would include the ethical maxims described in earlier chapters and the truth claims of religion we surveyed. It would mean, above all, that we are motivated by things other than those we see around us, of which, in religion, there are many—above all, God's glory.

All the same, one may suspect that even such other-worldly aspirations are merely this-worldly, since all of them are about things that ultimately find their death in this world—unless we should have some way of knowing that they have eternal value.

Alternatively, one may adopt an *interventionist* God who, through his being there, changes what we *expect*.[40] The universe is not closed. God reaches down into our reality. God *acts* in this world.

The world is not, therefore, all that it seems. This, too, is a central religious theme: "By You I can run upon a troop," wrote the psalmist David. "By my God I can leap over a wall."[41] We may expect the unexpected, through God.[42] The theologian Stephen H. Travis wrote that "to hope means to look forward expectantly for God's future activity."[43]

37. Camus, quoted in Schulz, *Critique of Hope*, 24.

38. This may apply as much to organizations as to individuals.

39. Rom 14:8.

40. Another possibility is revolt against a loss of meaning and purpose. Such revolt need not be rational or deliberate.

41. Ps 18:29.

42. Godfrey Vesey and Paul Foulkes note that the empirical prerequisite that leads us to the notion of god "lies in our coming up against powers that vastly exceed our own" (Vesey and Foulkes, "God," in *COLDP*, 127–28).

43. Travis, "Hope," 321. One could conceivably base hope on a social future. In the same way as I hope in God's future activity, I may hope in society's future activity. While we should not, in the words of philosopher and gender theorist Judith Butler, "accept a closing down of horizons," there are very many people for whom social hope

This de-objectifies hope. It *relativizes* it—because God's future activity cannot be known—and if God's future activity is always good, then, regardless, this translates fear and despair into hope.[44] In the words of Søren Kierkegaard, the only way to hope in this life is to "relate oneself expectantly to the possibility of the good." In fact, "at every moment always," he wrote, "one should hope all things."[45]

This provides hope even where we find no *reason* for hope. Hope in God offers a hope in the face of a world that does not satisfy—and will not satisfy a great many people within the span of one lifetime.[46]

Such hope is not defeatist. We continue to hold up the world in our minds to the world itself.

Paradoxically, if we accept the "God option" as the basis of all motivation, this would seem to be the option of deepest disillusionment, even despair—yet, at the same time, the option of greatest hope.[47] It would represent deepest disillusionment, because we understand that the world in our mind is no longer worth holding up to the world itself. It would represent greatest hope, because we find new motivation in God. This motivation, and the God who makes it possible, is *necessary* for us to carry on.[48]

Those of a supremely philosophical bent might ask, what if God is necessary for us to carry on, and yet he does not exist? We would, of course, have no reason to carry on. And yet we do.

---

will not be fulfilled (Gessen, "Judith Butler," q. 7). There is, too, a more existential possibility. Stephen H. Travis's "future activity" of God is surely God's present activity.

44. God's future activity must be extremely complex, if he intervenes in millions, if not billions, of people's lives. But that is if God acts individually with each one, as though he were composed of countless individuals. This may change, if he is One, everywhere and at all times.

45. Kierkegaard, *Works of Love*, 249. This may be stated more broadly: that God exists if there is a need for him. Traditionally, that need lies with the forgiveness of sins and, more recently, with the "ascent of man" (Bronowski, *Ascent of Man*, ch. 1, n.p.).

46. The philosophers Claudia Blöser and Titus Stahl note that "almost all major philosophers acknowledge that hope plays an important role in regard to human motivation" (Blöeser and Stahl, "Hope," introductory para.).

47. Paul Tillich wrote, "The acceptance of despair is in itself faith and on the boundary line of the courage to be . . . . The act of accepting meaninglessness in itself is a meaningful act. It is an act of faith" (Tillich, *Courage to Be*, 171).

48. A core question is whether we need a reason to carry on. Can we not just carry on? The fact that people do carry on may reflect a shallowness of thought rather than the ability to continue without reason. This contradicts our ethical maxim "Think expansively and holistically."

Chapter 40

Relations beyond Life

Our consciousness of death has much to do with our ability to trace relations beyond our own end. This we do in various ways.

WE HAVE SPOKEN OF ourselves as relation-tracing beings. All the time, we trace relations between things and other things: between low-level things and high-level things, things past and things future, things here and things there. But death, on the surface of it, implies the catastrophic severance of every relation in which we are involved.

On this, Simon Blackburn commented, "That might reasonably bother me a great deal."[1]

Blackburn was being conservative about it. The prospect of death does not merely bother us a great deal. In reality, the thought of death—particularly, imminent death—powerfully alters our thinking and behavior in various ways.[2] Socrates wrote, "People fear it, as if they knew that it is the greatest of evils."[3] It brings paralyzing fear, powerful denial, and radical reassessment of our lives. No one should consider themselves exempt from this. It can overcome any and all.

---

1. Blackburn, *Big Questions*, 192.

2. The writer and trader Daniel Defoe, writing on the plague, noted that the possibility of death "bring[s] us to see with differing eyes" (Defoe, *Journal of the Plague Year*, 174).

3. Plato, *Apology of Socrates*, para. 29.

The philosophical debate about death is most basically defined in terms of whether our present life is related to an afterlife or not. The operative word is "related." If it is related to an afterlife, we may ask on what *basis* this might be. If not, then we may ask what the absence of such a relation might imply.[4]

Relation-tracing is what makes us human. We all have the special ability to arrange our world, both conceptually and physically. In fact, it is our relation-tracing ability that enables us to transcend space and time and to pursue ambitions and aspirations that lie beyond the range of the rest of the animal kingdom—in fact, beyond life.

Excursus

But death is not a problem for all. Not all people fear death—and many are simply forgetful of such fear. What, then, shall we make of this?

On the one hand, we may accept the fear of death as a mere *phenomenon*—a fact of life wherever we find it—and, therefore, there may be little more to say. Perhaps, though, people *ought* to fear death.

If it is the case that they *ought* to fear death—more than that, fear it as a prelude to liberation from fear[5]—then it would seem correct to *introduce* in them the fear of death, together with *deliverance* from such fear. This would happen through the compassion described in earlier chapters.

Further, such relation-tracing has everything to do with *motivation*. Wherever we find that things are not so arranged as we think they *ought* to be, we are motivated to act. But if the arrangements in our mind have no reasonable prospect of fulfilment, this may ruin our motivation.

---

4. The distinction between a material body and an immaterial soul does not seem to be an issue here. Many non-dualists believe in immortality (Lewis, "Immortality and Dualism," in *Persons and Life*, 110).

5. This is of political significance, too. The philosopher Sidney Hook wrote, "Fear of death has been the greatest ally of tyranny past and present" (Hook, *Sidney Hook on Pragmatism*, 77). This link needs to be acknowledged before progress is possible.

Teilhard de Chardin wrote, "Man will never take a step in a direction he knows to be blocked."[6]

Plans and ambitions generally need to have some prospect of completion, or we do not undertake them—and death, it need hardly be noted, may rob us of such fulfilment. It makes a mockery of our goals. While death may not take away every motivation, it would seem to take away any *ultimate* motivation we may have.

The very fact that we resort to various (extreme) devices to sustain our life in the face of death indicates that death as the termination of life is a problem. More specifically, that the severance of relations is a problem.

Continuation

An important *aspect* of death is that its moment is fairly uncertain. As much as we might hope that we can control it, we do not know at what point it will intervene in our lives. Therefore, the arrangements that we make for the future—that is, relation-tracing—will inevitably be cut short by death at some point. We are not going to finish all that we began. In fact, the bigger the ambitions that we have, the more likely they are to be cut short by death.

Unless, that is, we could, in some way, continue our present story *after* death, in a *life* after death. There are various ways that this may conceivably happen:

We may continue our present *activities* after death: through hunting, say (the eternal hunting grounds are proverbial), through continued research, or by watching all the movies we ever wanted to see—even engaging in needlework or playing cards. While it is rarely assumed today that there will be a continuation of life in this form, one does still encounter the idea. More often than not, however, it is assumed that the story will continue in some other way.

It is a common idea that the real story of our life is one of sin and righteousness. Immortality is "the reward for righteousness and martyrdom."[7]

Or the real story of my life may be one of faith and apostasy. What would matter then is whether I had a saving faith or made a suitable *confession* of faith.

---

6. Teilhard de Chardin, *Phenomenon of Man*, 231.
7. K. Kauffman, "Immortality of the Soul," 6:564.

Or my eternal worth may be assessed in terms of my cultural or social status. What would matter then is whether I became worthy of honor, even veneration, by my people.

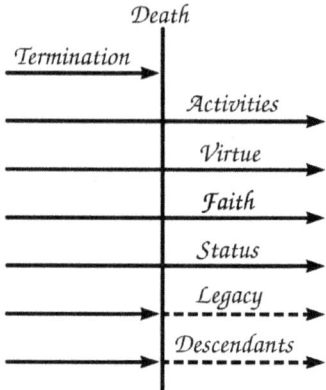

Superficially, the continuation of life after death—in whatever form—would seem to present us with a respectable answer to the puzzle of death.[8] With such prospects of continuation, we would retain our enthusiasm for the things of this life. Nothing need ever be lost—that is, nothing that really matters. It would not matter to us, therefore, if our hopes or plans should be cut short in this life.

However, there is an obvious difficulty here—for philosophers, at least. There is no evidence—not that we can agree on, anyway—as to whether there is continued consciousness after death.

> While I thought that I was learning how to
> live, I have been learning how to die.
>
> —LEONARDO DA VINCI

Yet even if there were no continuation of consciousness, there could be other ways to find some form of continuation.

Our activities in this life might leave a valuable legacy in this world. If we can see beyond our deaths, we can prepare a world for others—or,

---

8. Not all forms of continuation are equally desirable. In particular, some may justify behavior—say, suicide bombing—that we have ruled out in other parts of this metaphysics.

more selfishly, a world that exalts our own *contribution* towards others. We would, of course, assume—although it would seem presumptuous to some—that our legacy would be a good one.

Alternatively, we may live on through our children and our children's children, continuing as a *dynastic* presence, as it were.

And if we should want to be romantic about it, Edvard Munch wrote that we live on through the flowers that grow on our grave.[9]

Perhaps we should not think about this too closely. No legacy lasts forever. No family line is eternal. Some dynasties have been short, with brutal ends and no remembrance—while company fortunes have tended to be even shorter. In fact, even the stars will one day all go out.

Losing the Self

There may be another possibility. Perhaps we may lose our own self, even while we live.[10]

We may lose ourselves in our *society*—so that, to ourselves, we no longer exist. Our hopes, desires, and intentions would no longer belong to us. The notion of death is, after all, a very personal thing and would seem to be greatly exaggerated by our own self-awareness and self-importance. Might it not be possible, then, to blend with a collective stream of consciousness from generations past to generations future?

Yet any escape from our individuality—in the sense of entering into a self-effacing fusion with society—would seem to necessitate an exit from society as we know it.

Our society does not nurture selfless thinking. Society today is dependent throughout on individualism—on *my* wants, *my* intentions, *my* actions, *my* rights, *my* life. It is so diverse and fragmented that any attempt to reunite it in a fusion of histories, beliefs, and purposes seems impossible.

Or perhaps we would be able to lose ourselves in an infinite universe. "Forget yourself," wrote the artist Yayoi Kusama. "Become one with eternity."[11] Yet it seems unlikely that one should forget oneself in eternity, when one cannot forget oneself in society.

---

9. Or, less romantically, we may live on through donated organs.

10. More sinister is the possibility that we may lose our own self for a terrorist cause.

11. Kusama, "Forget yourself."

Then, there is the option simply of living in *tension*—in *terror*, for some—of the end of all our designs and dreams. This is, after all, what many people do with death. The only way to carry on, we might say, is to forge ahead with all our plans and ambitions, yet with *terror*—if we grasp the full reality of it. "Unreconciled" is the way that we should die, wrote Albert Camus.[12]

Perhaps there is a way to conquer death that the ancients could not fully have imagined. We may enter a phase of life, at the end of life, which we call *retirement*, in which there is nothing more to do, nothing left to lose. This is the time when we deliberately put it all behind us, burn our bridges, exit from the problems of humanity, and enjoy the afterglow. In this case, we have already died a certain kind of death—so that if we should die tonight, we might miss only a cup of coffee in the morning or a game of golf in the afternoon sun.

As tentative as this discussion may have been, we shall not leave the question of death and the afterlife without conclusion. We may only adequately understand death if it is the doorway to a continuation of life or *new* life—in a most *positive* sense.[13] In a negative sense, a continuation of life may offer us nothing over the termination of life—or worse.

Someone might object. "This is a fallacious appeal to consequences!" Yet in many cases, if we believe it because we *want* to believe it, our wanting to believe it may point to the *necessity* of such belief. In particular, we spoke in chapter 39 of the need for a hope that is not a false hope. Death may point to the partial nature of this life and a certainty that embraces both life and death.

---

12. Camus, *Myth of Sisyphus*, 55. There may be another kind of terror, namely, the terror of what may happen to us if we give up in the face of death.

13. This would be a hope against hope, as described in ch. 39.

Chapter 41

Meaning as Relations

Meaning is not merely about the ends and aims of my own existence. My own meaning is related to that of others.

WE TYPICALLY THINK OF meaning as meaning *for me*. Yet meaning is bound up with our wider context. My *own* meaning has a profound effect on others. That is, meaning is *social*. To think of it only in terms of my own, personal meaning is narrow—and not only narrow but perilous, as we shall see.

For many, the question of meaning is paramount—and a *lack* of meaning is no slight problem. Some experience absence of meaning as a living death, while others would rather die than surrender their meaning.[1] At the same time, there has been a curious *retreat* of meaning in our day.[2] It now lies beyond the interest of many people—even of dictionaries of philosophy. Yet historically, it has been an important philosophical theme.

What is meaning?

---

1. Rabindranath Tagore observed that people "completely identify themselves in their minds with their fixed surroundings and with whatever they have gathered, and to have to leave these is death for them" (Tagore, *English Writings*, 38). The philosopher and mathematician Hilary Putnam held that meaning is found only within a theory or description—implying that meaning is lost if this is destroyed (Devitt, "Shocking Idea about Meaning," 471).

2. The philosopher Karl Britton wrote in 1971, "The question of the meaning of life does not loom large in the teaching or writing or professional philosophers nowadays" (quoted in Seachris, *Exploring Meaning of Life*, 2).

There are various kinds of meaning: existential, psychological, linguistic, scientific, and more. The meaning that we explore here is the rather elusive "meaning of life." What is it that gives life its meaning?

The answers are many, yet it will often come down to one of three things:

- Have I found my place in the universe?
- Have I brought my own self to full expression?
- Am I achieving my purpose in life?[3]

We may summarize these three questions like this: I experience meaning when I know and feel my place in the big, wide whole—in the context of everything. This has everything to do with the low-level concept of the *relatedness* of all things and makes the question of meaning integral to a philosophy of relations. Meaning is about the way that I survey an endless expanse of relations—above all, whether I do so broadly or narrowly.

Intuitively, the meaning of life is *all-encompassing*. It is not *partial*. Aristotle wrote, "Happiness is the meaning and the purpose of life—the whole aim and end of human existence."[4] We may note his emphasis on the whole. There is nothing that falls outside of this meaning and purpose.

It has been the tragic tale of too many that they held a meaning that was later exposed as being only partial or was found to hold no real meaning at all. They found their meaning in causes that were destroyed, relationships that ended, even things that later proved to be ruinous both to themselves and to others.

However, not all philosophers agree. Many are willing to accept only *partial* meanings of life. "Big" meanings, in this view, are an illusion—and whenever we feel tempted to find them or think that we might be missing them, we should just not think on it too hard. Thomas Nagel wrote, "Perhaps the trick is to keep your eyes on what's in front of you."[5]

This may seem to have some appeal, except for two things. Firstly, we intuitively sense that meaning is not something we should lose. It ought to be timeless and absolute, not a victim of the vagaries of life.

---

3. These questions are derived from linguistic meaning. In fact, of whichever kind of meaning we speak, more or less the same questions apply.

4. Dobrin, "Happiness," para. 1.

5. Nagel, *What Does It All Mean?*, 99. This contravenes the ethical maxims we developed in chs. 8 and 9, above all expansive and holistic thinking. It is no category mistake to speak of meaning in this context.

Secondly—more than this—if our meaning is only partial or fleeting, we may bring others into bondage to our own meaning.[6]

Meaning Is More

On the surface of it, an all-encompassing meaning presupposes an all-encompassing system of thought—since it is only within a total system of thought, which includes all things, that meaning is found. If my own system of thought merely *appears* to be all-encompassing, there is the risk that I may stumble and fall outside it.[7]

Of course, if meaning is to be found in total systems, this contravenes our earlier maxim, *Avoid totalizing tendencies*. It contravenes other maxims, too. Clearly, there must be something else to meaning, if it is to be the kind of meaning we can embrace.

Consider the man who finds meaning in his rose garden. He wants nothing more, and nothing less, than this garden which he tends, day by day. All is well, while he measures his meaning by the well-being of the roses in his garden. Yet when he steps outside his garden, to ask what life is about beyond the roses, he finds no answer.

All is well, while I measure my meaning by the popularity of the music I make, the growth of my business, or my seat in parliament. Yet as soon as I step outside of such meaning, to ask what life is about *beyond* it, I find no answer—until I find a new and bigger answer that includes the old—or, if I can still believe it, *replaces* it.

> The system is a self-contained whole within
> which everything is made meaningful.
>
> —GRAHAM WARD

---

6. Or we may exclude the well-being of others. For this reason, I reject the Epicurean and Stoic idea of ataraxia, or peace of mind (Blackburn, "Ataraxia," in *ODP*, 27). While, in an important sense, this is desirable, it may keep us from the service of humankind. The philosopher and astronomer Pierre Gassendi, similarly, wrote that happiness consists in ease of body and peace of mind (Gassendi, "Peace of Mind," 13).

7. Total systems of thought are, however, beyond our grasp. Imagine, again, our universe as an endless expanse of relations. Since it is endless, it is beyond all systems. All of our systems represent mere regions of relations in the endless expanse.

It may be true to say that, as long as I find my meaning in my roses or the popularity of my music—and so on—I feel no *need* for anything else. Yet the awareness that such meaning can fail is always close by. Countless people have lived with suspicions—if not terrors—that their meanings could fail.

Whatever the case may be, we can always imagine meaning to be something *bigger*. The brand strategists Fabian Geyrhalter and David Glaze ask, "What is bigger than this?"[8] Meaning must be *more, more, more* than this.

As smaller meanings fail, and fail again, so we come to realize not only that meaning must be bigger but that, ultimately, meaning must be beyond all bounds. Ultimately, meaning is found in the endless expanse, beyond our own constructions, which confine it and reduce it to mere *regions* of relations. Ultimate meaning reveals itself to me only where *systems* of meaning are discarded.

Excursus

Does this mean, then, that meaning can be found—or that it cannot? Is meaning not lost, if the quest for a system of meaning is abandoned?

In terms of this metaphysics, meaning can indeed be found—at least, we can approach it ever more closely, as our horizons broaden in the context of "everything." However, it will not be found in its fullness in our own constructions, which confine it and reduce it.[9]

One has first to recognize the insufficiency of that by which one's meaning is now supported, before one relinquishes it to find a broader meaning. In this sense, one needs to abandon meaning before one finds it.

Presumably, this is why some of our meanings are more alluring than others. The larger meanings of this world *incorporate* our smaller meanings—which on their own fail to answer the search for "more."

---

8. Geyrhalter and Glaze, *Bigger Than This*, back cover. This is written in the context of creating "meaningful" branding.

9. One might refer to such constructions as idols.

Larger meanings are systems outside of the confines of our own small, muddling purposes.

This is why a Napoleon, a Hitler, or a Pol Pot can exist. They offer us "more." So much more, in fact, that it may be difficult to imagine anything outside or beyond the ideologies they represent. Sometimes, not even famine, war, or grievous sanctions and blockades may defeat such systems—so overpowering may they become.

If our meaning is not to be found in the totality of the endless expanse—or in something that transcends all other meanings—then our meanings may well lead to disaster. History has shown us this all too often.

What, then, of *God*? For many, there is no greater meaning to be found than in God—yet many *atrocities* have been committed in God's name, not to speak of injustices and injuries.

However, we may understand God not as one who justifies my own system of meaning—"I was obeying God," or "God told me to"—but as the purpose beyond all purposes, before whom all contrived meanings dissolve. Paul Tillich wrote about "God above God," who is being itself, not a being.[10] We may interpret this as belief in a God who does not serve our own small designs.

If God should be the guarantor of meaning, this would be so not because we find our meaning in doing something big enough or complete enough for God. No action can ultimately be big enough or complete enough to secure my personal meaning. Rather, God would guarantee my meaning through that which he gives me to do, even if it may seem small or ordinary.

The Loss of Meaning

Above all, the problem of a loss of meaning is not merely my own. It may reduce the meaning of *others*. One is able to view others only in terms of the relations one has traced oneself in this world. There are no other relations that one traces. This means that, for me, everything must be related to my own personal world of meaning.

When a woman embarks on studies in sociology, she may be heard to say, "*Sociology* explains all that." When a businessman owns a nature reserve, he will run it as a business. If an economist becomes a president, he

---

10. Tillich, *Courage to Be*, 176.

will treat the nation as an economy. One president complained that a fatal pandemic was bad for the economy. He had been trained as an economist.

The earth quakes, wrote King Solomon, when a slave becomes king—presumably because the slave sees the world in terms of the master-slave relationship and cannot see its meaning beyond such terms.

It is only when we find our meaning in *everything*—not through rationally constructed meanings, which are partial—that we approach true meaning, which is all-encompassing.[11] Meaning is not found in reason but in wonder.[12]

Finally, the fact that "meaning" has often lost its relevance today may have much to do with the fragmentation of society. With ever more complex diversity and specialization, in all areas of life, we have lost touch both with our world as a whole and with one another. More and more, we have needed to focus on smaller worlds—and meanings. When these smaller meanings begin to dissolve, we seek to preserve them. If they represent a system for me that I cannot see beyond, I may choose any means at all to preserve them—among them, fraud, lies, threats, violence, even worlds of fancy in my mind.

---

11. Stephen Hawking wrote, "If everything in the universe depends on everything else in a fundamental way, it might be impossible to get close to a full solution by investigating parts of the problem in isolation" (Hawking, *Brief History of Time*, 12). This applies, too, to meaning.

12. In a sense, there is no merit in finding meaning for our lives. It lies beyond the best powers of reason and morality.

Chapter 42

What Is Metaphysics?

*In broad compass, we survey how this metaphysics
has shifted the contours of metaphysics.*

BROADER IN SCOPE AND more fundamental than science, metaphysics seeks to understand *everything, briefly*.[1] Many have said that, in principle, it cannot be done. David Hume wrote that every metaphysics should be committed to the flames, "for it can contain nothing but sophistry and illusion."[2]

Hume, however, created his own problems, by sundering things from relations, and fact from value—which, in later chapters of this book, we have both dissolved and fused. Small wonder that he blamed it on philosophy.[3]

At the same time, we may be thankful that metaphysicians and philosophers like Hume, together with many others in many fields, changed the way we see things and fundamentally changed our world—often at great effort and pains. They pushed us to the limits of thought and set our fictions aside. More than this, they brought us incrementally to the point

---

1. Relatively speaking. Some metaphysics are very large—yet small compared with what they gather under one scheme of thought.

2. Hume, *Enquiry*, 120.

3. The mathematician and philosopher Alfred North Whitehead said, "Every scientific man in order to preserve his reputation has to say he dislikes metaphysics. What he means is he dislikes having his metaphysics criticized" (quoted in Conger, "Whitehead Lecture Notes").

where we see an *interconnected* world, in a way we never saw it before. It is on this basis that we have been able to suggest a way forward.

Yet as we create something new, the purpose is not to *overthrow* what has gone before but to unify and reconcile: fact and value, science and mind, mind and body, reason and passion, free will and determinism, religion and reason, the individual and the group, modernism and postmodernism—even Eastern and Western.[4] In the spirit of Clement of Alexandria, "What I mean by philosophy is not Stoicism, nor Platonism, nor Epicureanism, nor Aristotelianism, but the sum total of all the good things said by these schools in the teaching of justice and truth."[5] The aim is, as far as possible, a Grand Unified Philosophy. "To understand," wrote Camus, "is above all to unify."[6]

The purpose of this book is not, therefore, in the custom of the philosophers, to supplant previous thinking but, as far as possible, to reconcile divergent views and to incorporate them in a systematic, harmonious whole.

Changed Metaphysics

In the course of this metaphysics, some things have changed about metaphysics in general. In some cases, without announcing them, we have done a number of things.

Firstly, we have revisited primordial ideas. We have turned to concepts that are far more basic than those which are normally employed in philosophical schemes. We have called these low-level concepts. They infiltrate our thinking everywhere. They are present in all of our speaking and writing.

It is a mistake to think on philosophy on a higher level, in the language of more common concepts, such as realism and anti-realism, theism and atheism, altruism and egoism. It is a mistake because of the complexity of these terms and the bondage of our minds. These concepts lead to conflict and contradiction, and obstruct the path to harmony, integration, and wholeness.

---

4. Richard Osborne pondered, "If there is a global economy—and a world culture—can there be a 'world philosophy'? Is it possible to amalgamate Eastern mysticism and Western rationalism in a new philosophical synthesis?" (Osborne, *Philosophy for Beginners*, 172). The answer here is yes, it is possible.

5. Daniel-Rops, *History of the Church*, 1:344.

6. Camus, *Myth of Sisyphus*, 22.

One may imagine a windmill in the desert, pumping water from the ground, pouring water into a tank. Suddenly, the water stops flowing. A farm is imperiled. The farmer checks his rain gauge. He checks the wind speed. The rainfall has been normal, the wind speed has been constant. He worries about the water table. Perhaps the neighbor's new windmill has caused it to drop.

Not finding the problem in any of these things, his thoughts bring him to the windmill itself. As he goes over it, part by part, he finds a single bolt, which has snapped off and fallen into the sand. He picks up a broken piece and takes it to find a replacement. The problem lay at a simple, low level.

This is not to say that high-level concepts are not useful. Their use throughout this metaphysics is proof enough that they have great value in formulating and exploring ideas. However, they have failed us in delivering the answers we need.

Secondly, this metaphysics separates itself from previous metaphysics in that we do not set up another stable center, starting point, or single thought—or anything of that sort. We understand that the relations that we trace are finite, as our minds are finite—in a world that presents us with *boundless* relations between things. This rules out the possibility of planting our flag, so to speak, in one place rather than another, as was done before.

If a man will begin with certainties, he shall
end in doubts; but if he will be content to begin
with doubts, he shall end in certainties.

—FRANCIS BACON

The proposal here is not a new metaphysics in the same way as older metaphysics. Nothing is left of single thoughts: the categorical imperative, absolute idealism, the will to power, manifest destiny, a proletarian dictatorship, apartheid, or any of the many labels one may put on first philosophies of the past.

Rather, relations between things offer us a center *without* a center—a philosophical theory of relativity—a description of everything, briefly, which *removes* the central idea, leaving behind only relations between things. In order to overcome the relativity of thought, it embraces it, by beginning with the whole, not the parts.

Thirdly, since Hume, and more particularly since Wittgenstein, it has been hard to see how a metaphysics should speak about *ethics*. Wittgenstein claimed to have run up against the limits of language, which left him without the password to advance to ethics. He was famously reduced to statements such as "I know that that's a tree," and then, "We are only doing philosophy."[7]

The exclusion of value judgments from rational philosophy has had a profound effect on metaphysics and on our world. We encounter widespread disorientation, on the one hand, and a yearning for direction, on the other, while new ethical bearings seem nowhere to be found.

This metaphysics has sought once more to incorporate ethics into metaphysics, turning to *emergence* for a solution. On the basis of the broader features of language and reality, we find moral direction.

Fourthly, *religion* is a theme that has accompanied this metaphysics through several chapters. Religion has a major, if not dominant, influence on our world and always has had. It deserves to be dealt with for the weight of its effects alone.

Yet metaphysics has all but bypassed major concepts of religion today. While there always will be a line between metaphysics and religion, the proposal here is that we can know enough about religion, through metaphysics, to bring about an intelligent meeting of the two.

Among other things, our thoughts on religion have presented us with concepts about God, awakening, religious truth claims, hope, and various ideas besides, which serve to align religion with philosophy.

While we have sought not to stray across the outer perimeter of philosophy, there may be lines of thought that could be picked up by theologians at the other side.

Fifthly, we have effectively set aside the classical questions of *being*, which have often dominated metaphysics of the past. The *being* of a thing, philosophers said, is what makes it an entity. An entity thus becomes an item in a logical construction.[8] For thousands of years, this has

---

7. Wittgenstein, *On Certainty*, 467. Wittgenstein brought a wealth of philosophical background to such statements so that they are not senseless. The point here is that such thinking excluded ethics.

8. This reflects on the subject of *logic*, which, although important to philosophy, has not been singled out in this book. Logic is about the rules by which inferences are drawn—which, again, is about how we relate things to things.

What Is Metaphysics? 301

represented one of the silent forces of philosophy—the elephant in the room—the one conspicuous error we overlooked.

Here we conclude that there are no things and no relations—no fact and no value—no cause and no effect, no objects, no entities, no events. They are all illusory. While, when we speak of them in haste, they seem to be real to us, when we examine them more closely, they vanish into an endless expanse.

Vanish? What, then, is left of this philosophy?

However, we have said that, in a sense, the illusion is real. We gave the example of a glass marble, which we tap on a tabletop. While this marble is "mostly void," we *experience* it to be hard and round. Our everyday experience of things is largely unchanged—yet we have needed to examine these things closely (and, at first sight, strangely) to discover solutions to major philosophical problems.

Paradoxcially, at the same time as we have deconstructed everything, we have sought to restore some reality to abstract words such as destination, ownership, citizenship—ultimately love, beauty, meaning—words that are born of relations between things and, in the case of symbolic language, encompass *expansive* regions of relations.

Sixthly, we have reduced *science* to an equal among equals with other modes of thought. This was not merely a semantic trick. We showed how science represents an imperfect fit with reality; in fact, fails in a global system; and, finally, merely obeys the "purpose of reason," which is to bridge the gaps that we find, whatever these may arbitrarily be.

We see the old view of science in the philosopher Auguste Comte.[9] There are three major stages, he wrote, in the development of the human mind:

- the theological,
- the metaphysical, and
- the positive. Here, in this third and highest stage of development, superstition and metaphysics give way to science.

Consider how the pendulum has swung. We have leveled science to the level of value judgments, as all things are value judgments. Science, both in what it does and does not say, is subjective.

Further, we have leveled it to a level of abstraction among abstractions that include our natural, everyday language.

9. Comte, *Course in Positive Philosophy*, 6.

Part VI: The Ultimate

No longer do we have reason to privilege the natural sciences. In fact, the very opposite. It is now the *human* sciences—the uncertain sciences, so-called—that present the greater and more important challenge for human knowledge and wisdom.

Excursus

It would seem a good place to ask: in what way, then, is this philosophy *postmodern*? Are not postmodernism and metaphysics a contradiction in terms? While this book does not in every way bear the marks of postmodernism, it is postmodern in fundamental respects.

It rejects the "single thought" of Schopenhauer,[10] which is found in various modern and premodern philosophers. In its place are things and relations, which in themselves have no central idea. In fact, things and relations are set aside in the later chapters of this book and, with them, everything we derive from them.

Everything, Briefly undermines modern metanarratives—above all, the supremacy of reason. It does so by diminishing the "purpose of reason," then dismantling the low-level ideas on which it rests. In keeping with this, it rejects the narrative that scientific thinking is either objective or progressive. Science is subjective—in fact, so subversively subjective that we have been unable to recognize the fact.

Finally, based on a survey of things and relations, *information* becomes a critical aspect of our world, since there can be neither things nor relations without it. With an emphasis on information, we dismantle what a senior editor of *Encyclopaedia Britannica*, Brian Duignan, called "the unjust hegemony of Enlightenment discourses."[11] In their place, we have proposed equality of information and the necessity of its uptake in various systems.

---

10. Schopenhauer, *World as Will*, preface. Blaise Pascal called it "everything contained in one word" (Pascal, *Thoughts*, 259).

11. Duignan, "Postmodernism and Relativism," para. 4.

The End

Some have said that a new metaphysics is impossible, by reason of the explosion of knowledge today. It becomes increasingly difficult to master even small corners of our knowledge and understanding. It may take years to gain a thorough knowledge of the most basic things: geometry, boat-building, or poetry. How, then, can a metaphysics remotely hope to encompass the vastness of our world?

Yet as in the past, so in the present. The human race has had an extraordinary ability to sift and sort knowledge, in such a way that all of its most important aspects are open to view: in dictionaries and encyclopaedias, and, more recently, through subject encyclopaedias and search engines.

This metaphysics has been written, then, in the belief that we have a sufficient grasp of reality and sufficient expertise to *interpret* the knowledge that we have.

At the same time, we must end with a confession. In the very act of writing it, this metaphysics has gradually been overthrown. We come to the troubling realization that it has been superseded by its own observations. Not only have we thrown away the ladder that we climbed; we have found that there was no ladder at all.

We started out by saying that all things are related to all things and that, from this, we may derive conclusions that are "clearly and distinctly perceived." On this basis, we distinguished between balanced and broad thinking, on the one hand, and parochial, short-sighted, or self-interested thinking, on the other. The way that we trace relations, we said, may make us wise or make us fools. It may redeem the world or destroy it.

The truth is that *all* the relations that we trace are deeply inadequate. All are unsound. All are deeply corrupted. In fact, at the end of the day, there is not a thing that coheres, so that nothing can be related, but everything dissipates into the farthest reaches of the universe.

The poet John Dryden asked,

> How can the less the greater comprehend?
> Or finite reason reach infinity?[12]

There is no way in which we may validly trace relations between things. These things and the relations between them are, at the end of the day, a reduction or selection of all things and relations. Max Black wrote that

---

12. Dryden, *Poetical Works*, 1:162.

our very language "necessarily distorts and misrepresents"[13]—and as we have described it in this book, the very words that we have inherited from our forebears are often filled with destruction within.[14] Every tracing of relations, therefore, is finally doomed to defeat.

Yet there is encouragement even in this. As we come to understand it, we dissolve the notions that have tended to make us reductive, disruptive, and obstructive, and we become more humble, more human, and more free.

The words of the Roman missal of 1570 come to mind:[15]

> *Confiteor Deo omnipotenti . . .*
> *quia peccavi nimis cogitatiine, verbo et opere:*
> *mea culpa, mea culpa, mea maxima culpa.*
>
> I confess to almighty God . . .
> that I have sinned exceedingly in thought, word, and deed:
> through my fault, through my fault, through my most grievous fault.

---

13. Black, *Labyrinth of Language*, 160.

14. A recurrent theme in Blaise Pascal's writing is the wretchedness of human nature—and Albert Camus wrote, "If the only significant history of human thought were to be written, it would have to be the history of its successive regrets and its impotences" (Camus, *Myth of Sisyphus*, 18).

15. Lambert, "Compline," in *Vespers-Book*, 59.

Glossary

EVERYWHERE, THE DEFINITIONS OF words vary. Rather than seeking *ideal* definitions, this glossary seeks to provide basic definitions of key words in this book, to indicate how they are used.

abandonment A deconstructing and reordering of every relation one has ever traced. A revaluing of all values.

anaphora A linguistic feature, usually thought of as linking two sentences *explicitly*.

atomic word The view that a word has no subatomic existence. Leonard Bloomfield's "minimum free forms." *See also* **subatomic**.

awakening A fundamental decision, through which we are willing to abandon every relation we have ever traced.

balance To relate things one to another so as to present a proportionate and harmonious whole.

bondage Our inability to escape our own thoughts or actions.

boundless relations Innumerable relations between things. John Locke's "almost infinite."

bridging inference A linguistic feature that ties two sentences together implicitly, through the things we *infer* about words.

built environment Areas of human activity or management.

calling The union of all that a person is with their employment.

coercive institutions The instruments of state.

concept Identical with a thing. *See also* **thing**.

connotation The illusion of denotation. *See also* **denotation**.

continuation The continuation, in some form, of our present life in an afterlife.

critical theory In terms of this work, theory that resists the suppression of information or the prevention of its uptake.

death The end of the life of a person or dynamic system of organization.

democracy A system of government in which all eligible citizens participate in political decision-making.

denotation The dictionary definition of a word. *See also* connotation.

dissipated thing The view that the identity of a thing dissipates throughout an endless expanse of relations. *See also* **dissipated word**.

dissipated word The view that the identity of a word dissipates throughout an endless expanse of relations. A further development of the subatomic word. *See also* dissipated thing.

economics The production, consumption, and transfer of wealth—driven lately by algorithms.

education Imparting the ability to trace relations, above all as a balanced and harmonious whole.

endless expanse A boundless (almost infinite) *expanse* of relations.

equality benefits Where absolute rights seem irreconcilable, benefits take the place of penalties.

equation A statement that the values of two mathematical expressions are equal, before one has matched them up with reality.

esthetics A tracing of relations that brings things into balance, typically in a non-utilitarian way.

ethics A set of moral principles or dispositions that represent a balance of relations. *See also* **balance**.

event A thing that happens. *See also* **thing**.

fact A statement about relations between things, which may be affirmed or denied, and used as the basis of argument.

falsifiability The possibility of falsifying statements or theories in a closed system.

first philosophy Another term for a metaphysics, especially one's personal metaphysics. *See also* **metaphysics**.

foundation The universal context in which thought is embedded.

free will The power of acting without the constraint of necessity or fate.

God A personal, supreme being.

grammar A method of compressing the relations we trace between things. The method varies, depending on the need.

harm Behaviors individual or collective, which may be heritable, that impose injury, suffering, distress, and destruction on our world.

high-level A relative term. Concepts which lie at a level of abstraction above that which is normal or average. *See also* **low-level**.

history Relations between past events.

holism The theory that all the parts of a whole are intimately interconnected.

hubris Excessive pride or self-confidence, especially as it relates to knowledge.

idea Something less exact than a concept. *See also* **concept**.

identity Identity is ordinary. It is found in the way in which individuals relate things to things.

immutability The general inability of humans to change the relations that they trace or to change them quickly.

individuation To distinguish a thing from everything else, whether it be an object, event, or concept.

infinity That which is in principle so numerous or large that it is impossible to measure or calculate.

information An objective, mind-independent entity that can be encoded and transmitted.

instruments of state Coercive insitutions that safeguard a broad balance in society.

invalidation The falsification of statements or theories in a global system. *See also* **falsification**.

is–ought distinction The distinction between the way things are related and the way things ought to be related.

language Methods of human communication that convey the relations between things.

linguistic system A system through which we arrange our language, to a large extent of our own choosing.

low-level A relative term. Concepts that lie below a level of abstraction that is normal or average.

mathematics With the exception of pure mathematics, the reduction of things and relations to a high level of abstraction.

maxim A short statement of a general rule of conduct.

meaning To know and to feel one's place in the whole, in the context of everything.

meta-feature A distinctive attribute of the endless expanse of relations, at an emergent level. *See also* **endless expanse**.

metaphysics A grand unified philosophy, which integrates—or provides a *means* to integrate—all fields of knowledge and experience using familiar philosophical concepts.

mind–body dualism The notion that an immaterial mind interacts with a material body.

moral Concerned with right and wrong behavior.

motivation That which moves us to action.

natural environment The phenomena of the physical world, beyond the built environment, though not unaffected by it. *See also* **built environment**.

negative ethics What we ought *not* to do, as opposed to what we *ought* to do. *See also* **positive ethics**.

ontology The subject that deals with the nature of being and the kinds of things that exist.

ownership To come to terms with who one really is. To set aside pretences and illusions.

philosophy of relations To see that everything is intimately present to everything else and to draw consequences from the same.

positive ethics What we *ought* to do, as opposed to what we ought *not* to do. *See also* negative ethics.

positive religion According to Georg W. F. Hegel, an objective system of rules and laws, dictated by people or gods.

Glossary

pseudo ethics Ethics that is purportedly based on fact.

real That which largely corresponds with the reality we perceive. *See also* **reality**.

reality The state of things as they actually exist. In terms of this work, largely accepted at face value.

reason A flagging of contradictions, which are givens, and the thought process through which we seek to reconcile them.

region of relations Webs of relations we trace, yet significantly less than the endless expanse. *See also* **endless expanse**.

relations Relations between things. In terms of this work, such relations may not exist, yet we refer to them for practical purposes.

religion The belief in a superhuman controlling power, especially a personal God or gods.

science The body of knowledge accumulated and brought under general principles through the scientific method. *See also* **scientific method**.

scientific method The systematic acquisition of knowledge through the application of rigorous principles and procedures.

self-assembly Our natural habit of assembling systems, especially as it relates to metaphysics. *See also* **unacknowledged metaphysics**.

structural linguistics The broader definition that applies to a cluster of theories. A complete, self-contained, self-regulating linguistic system.

subatomic word The view that a word has a subatomic existence, containing a profusion of relations inside. *See also* **atomic word**.

supernatural A transgression of the laws of nature, or any intervention of God in this world.

symbol Any thing, among other things objects and words, that stands in a relationship of metaphor to another. *See also* **symbolism**.

symbolism A relationship of metaphor: x stands for y. *See also* **symbol**.

synessence A philosophical term. The essence of a thing, which includes every possible aspect of that thing. A reversal of the usual meaning of essence.

synotation A linguistic term. All personal knowledge contained within a word. Every possible association or relation of that word.

syntocracy A form of government in which supreme power is vested in relations, through the people.

technology The application of scientific knowledge for practical purposes, especially in industry.

theory In this book, an equation or equations that have been applied to reality (but not all theory requires equations).

thing An object (a material thing), an event (a thing that happens), or a concept (a thing that exists in our mind). The contents of a word.

things-relations distinction A distinction between things and how we relate them. In terms of this work, it may not exist.

totality The whole of something, especially all relations.

tracing relations Marking the course of relations. Pursuing them as we connect one thing with the next and the next.

truth The final destination of philosophy, whatever this would be.

truth claims Propositions that can everywhere, at all times, be judged to be true. Core beliefs with philosophical foundations. *See also* **foundation**.

unacknowledged metaphysics A metaphysics embedded in one's language, usually unconsciously, which takes precedence over our conscious metaphysics.

unvirtuous behavior Behavior that contradicts virtue in its widest sense. Includes foolishness and fallacy, lies and deceit, and various forms of violence.

value That which I consider ought to happen, or which one ought to do. The evaluation of things or statements of the same. *See also* **fact**.

weighted word A word, the synotation of which overlaps more often than usual with the synotations of other words. *See also* **synotation**.

word A wrapper for a world of relations inside. A label of a thing. *See also* **thing** and **relations**.

worldview The way we arrange the world in our minds.

Bibliography

THE WORKS CITED ARE at a variety of levels. Some are designed for the general reader, while others require a close familiarity with philosophy and related subjects.

Aho, James. *Confession and Bookkeeping: The Religious, Moral, and Rhetorical Roots of Modern Accounting.* New York: State University of New York Press, 2005.
Aizawa, Ken. "Productivity of Thought." Dictionary of Philosophy of Mind, last updated May 2004. https://sites.google.com/site/minddict/productivity.
Allaby, Michael, ed. *A Dictionary of Ecology.* Oxford, UK: Oxford University Press, 2010.
Allard, James W. "Bosanquet, Bernard." In *CDP*, 82–83.
Alexander, Patricia A., and Philip H. Winne, eds. *Handbook of Educational Psychology.* 2nd ed. New York: Routledge, 2012.
Althaus, Horst. *Hegel: An Intellectual Biography.* Hoboken, NJ: Wiley, 2016.
Ambedkar, Bhimrao R. *The Buddha and His Dhamma.* Edited by Frances W. Pritchett. New York: Columbia University Press, 2021. http://www.columbia.edu/itc/mealac/pritchett/00ambedkar/ambedkar_buddha/01_5.html.
"Aristotle." Northampton, UK: Northampton University, 2021. https://www.northampton.edu/documents/Subsites/HaroldWeiss/Existentialism/Aristotle%20-%201st%20two%20pages.pdf.
Aristotle. *The Metaphysics.* Translated by John H. MacMahon. London: Bell and Sons, 1896.
———. *The Organon, or Logical Treatises with the Introduction of Porphyry.* Translated by Octavius F. Owen. 2 vols. London: Forgotten, 2017.
Armstrong, David M. *A Materialist Theory of the Mind.* London: Routledge and Kegan Paul, 1968.
Attenborough, David. "Prince William Interviews David Attenborough at Davos." YouTube, Jan. 22 2019. https://www.youtube.com/watch?v=0UIp7YvfOoI.
Aurelio, Marco. "The 70 Most Famous Phrases of Marco Aurelio." *Psychology.* Warbleton, UK: Warbleton Council, 2021. https://en.nsp-ie.org/frases-marco-aurelio-6305.
Ayer, A. J. "A. J. Ayer on Logical Positivism and Its Legacy." YouTube, Nov. 8 2015; orig. *BBC*, 1976. https://www.youtube.com/watch?v=nGoEWNezFl4.

Bacon, Francis. *The New Organon*. Edited by Lisa Jardine and Michael Silverthorne. Cambridge Texts in the History of Philosophy. Cambridge: Cambridge University Press, 2004.

———. *The Works of Francis Bacon*. Edited by James Spedding et al. 15 vols. Farmington Hills, MI: Gale, 2010.

Bailey, Richard. "Healthy Body and a Sound Mind?" Psychology Today, July 2, 2014. https://www.psychologytoday.com/us/blog/smart-moves/201407/healthy-body-and-sound-mind.

Bain, Alexander. *Deduction*. Vol. 1 of *Logic*. New York: Longmans Green, 1870.

Barad, Karen. "Karen Barad: Re-membering Time. For the Time Being." YouTube, Mar. 31 2021; orig., Temporal Belongings Research Network, Mar. 15, 2021. https://www.youtube.com/watch?v=MpD_N2FnV5M.

———. *Meeting the Universe Halfway: Quantum Physics and the Entanglement of Matter and Meaning*. Durham, NC: Duke University Press, 2007.

Barnett, Lincoln K. *The Treasure of Our Tongue: The Story of English from Its Obscure Beginnings to Its Present Eminence as the Most Widely Spoken Language*. New York: Knopf, 1964.

Barrow, John D. *Impossibility: The Limits of Science and the Science of Limits*. Oxford, UK: Oxford University Press, 1998.

———. *New Theories of Everything: The Quest for Ultimate Explanation*. Oxford, UK: Oxford University Press, 2008.

Barsalou, Lawrence. "Context-Independent and Context-Dependent Information in Concepts." *Memory and Cognition* 10 (1982) 82–93.

Baumer, Franklin Le Van. *Modern European Thought: Continuity and Change in Ideas, 1600–1950*. London: Macmillan, 1977.

Beard, Charles A. "Annual Address of the President of the American Historical Association, Delivered at Urbana." *American Historical Review* 39 (1933) 219–31.

———. "Written History as an Act of Faith. Annual Address of the President of the American Historical Association, Delivered at Urbana. December 28, 1933." *American Historical Review* 39 (1933) 219–231.

Beard, Matthew. "Courage Isn't about Facing Our Fears, It's about Facing Ourselves." Ethics Centre, Aug. 22, 2019. https://ethics.org.au/courage-isnt-about-facing-our-fears-its-about-facing-ourselves/.

Bedürftig, Thomas, and Roman Murawski. *Philosophy of Mathematics*. Berlin: De Gruyter, 2018.

Bell, Daniel. *The Coming of Post-Industrial Society: A Venture in Social Forecasting*. New York: Basic, 1976.

Bense, Hermann. "About Predicator.Name." Ontology4.us, Apr. 2014. https://predicator.name.

Bentham, Jeremy. *The Works of Jeremy Bentham*. Edited by John Bowring. 11 vols. Edinburgh: Tait, 1838.

Bentley, Madison. "The Psychology of Mental Arrangement." *American Journal of Psychology* 13 (1902) 269–93.

Bergland, Christopher. *The Athlete's Way: Training Your Mind and Body to Experience the Joy of Exercise*. New York: St. Martin's, 2008.

Bergson, Henri. *Creative Evolution*. Translated by Arthur Mitchell. New York: Holt, 1911.

Berkeley, George. *A Treatise Concerning the Principles of Human Knowledge*. Edited by Thomas J. McCormack. Mineola, NY: Dover Philosophical Classics, 2019.

Berlin, Isaiah. "Berlin, Sir Isaiah." In *Dictionary of Philosophy*, edited by Thomas Mautner, 67–69. London: Penguin, 2000.

———. "Isaiah Berlin Interview on Why Philosophy Matters." YouTube, Sept. 17, 2017; orig. *BBC*, 1976. https://www.youtube.com/watch?v=vib2rqJKSo8.

Berlinghoff, William P. *A Mathematics Sampler: Topics for the Liberal Arts*. Lanham, MD: Rowman and Littlefield, 2001.

Berry, Thomas. *The Dream of the Earth*. Berkeley, CA: Sierra Club, 1988.

Bett, Richard, and Donald R. Morrison. "Socratic Ignorance." In *CDP*, 215–36.

Bhalerao, Kinnari. "Interacting with Dr. R. Chidambaram." IEEE-VIT, Apr 14, 2021. https://ieee.vit.edu.in/ieeevitblog/vit/iris%202021/2021/04/14/Interacting-with-Dr.-R.-Chidambaram.html.

Biase, Luca de. "A Science of Consequences." In *Know This: Today's Most Interesting and Important Scientific Ideas, Discoveries, and Developments*, edited by John Brockman, 499–502. New York: Harper Perennial, 2017.

Biddiss, Michael D. *The Age of the Masses: Ideas and Society in Europe since 1870*. London: Penguin, 1977.

Bidny, Sophia. "The History of Zero." Kobotis Technologies, Sept. 17, 2014. http://kobotis.net/math/MathematicalWorlds/Fall2014/131/Presentations/pdf/BidnyS_p1.pdf.

Black, Max. *The Labyrinth of Language*. London: Penguin, 1972.

———. *The Prevalence of Humbug and Other Essays*. Ithaca, NY: Cornell University Press, 1983.

Blackburn, Simon. *The Big Questions: Philosophy*. London: Hachette, 2012.

———. *The Oxford Dictionary of Philosophy*. Oxford, UK: Oxford University Press, 1996.

———. *Truth: A Guide for the Perplexed*. London: Penguin, 2006.

Blake, William. *The Complete Poetry and Prose of William Blake*. Edited by David Erdman. Oakland: University of California, 2008.

Blöser, Claudia, and Titus Stahl. "Hope." In *SEP*, Mar. 8, 2017. https://plato.stanford.edu/entries/hope/.

Bloom, Benjamin S. *Taxonomy of Educational Objectives: The Classification of Educational Goals*. Philadelphia: McKay, 1972.

Bloomfield, Leonard. *Language*. Chicago: University of Chicago Press, 1984.

Blum, Fred H. "Max Weber's Postulate of 'Freedom' From Value Judgements." *American Journal of Sociology* 50 (1944) 46–52.

Bohman, James. "Critcal Theory." In *SEP*, Mar. 8, 2005. https://plato.stanford.edu/entries/critical-theory/.

Bonhoeffer, Dietrich. *Prisoner for God: Letters and Papers from Prison*. Edited by Eberhard Bethge. Translated by Reginald H. Fuller. New York: Macmillan, 1959.

Booth, Gordon T. *Evangelical and Congregational: The Principles of the Congregational Independents*. Beverley, UK: Evangelical Fellowship of Congregational Churches, 1995.

Bosanquet, Bernard. *The Philosophical Theory of the State*. Kitchener, ON: Batoche, 2001.

Bowles, Samuel, and Herbert Gintis. "Homo Reciprocans." *Nature* 415 (2002) 125–27.

Box, George E. P. "Science and Statistics." *Journal of the American Statistical Association* 71, no. 356 (Dec. 1976) 792–99.

Bradley, Francis H. *Appearance and Reality: A Metaphysical Essay*. Oxford, UK: Clarendon, 1966.

Bragg, Melvyn, presider. "Human Nature Is Better Understood through Science: Royal Institute of Philosophy Debate 2015." YouTube, Jan. 8, 2016; orig. Royal Institute of Philosophy, Nov. 16, 2015. https://www.youtube.com/watch?v=BVtfTXdURnc.

Brandl, Simon J. "Which Species Should We Save from Extinction?" World Economic Forum, Jan. 12, 2015. https://www.weforum.org/agenda/2015/01/which-species-should-we-save-from-extinction/.

Bridgman, Percy Williams. *The Logic of Modern Physics*. London: Macmillan, 1958.

Brittain, Charles, and Peter Osorio. "Arcesilaus." In *SEP*, Jan. 14, 2015; revised July 2, 2021. https://plato.stanford.edu/entries/arcesilaus/.

Bronowski, Jacob. *The Ascent of Man*. London: MacDonald Futura, 1973.

Broome, John. *Rationality through Reasoning*. Hoboken, NJ: Wiley Blackwell, 2013.

Browne, Thomas. *The Works of Sir Thomas Browne*. Edited by Grant Richards. 6 vols. Norderstedt, Germ.: Books on Demand, 2020.

Brunner, Emil. *Man in Revolt: A Christian Anthropology*. Translated by Olive Wyon. Cambridge, UK: Lutterworth, 2002.

Buber, Martin. *I and Thou*. Translated by Ronald G. Smith. Edinburgh: T&T Clarke, 1984.

Burg, David F. *A World History of Tax Rebellions: An Encyclopedia of Tax Rebels, Revolts, and Riots from Antiquity to the Present*. Abingdon-on-Thames, UK: Taylor & Francis, 2004.

Burke, Peter J. "Identity Change." *Social Psychology Quarterly* 69, no. 1 (2006) 81–96.

Burkhard, John J. *Apostolicity Then and Now: An Ecumenical Church in a Postmodern World*. Collegeville, MN: Liturgical, 2004.

Camus, Albert. *The Myth of Sisyphus*. Translated by Justin O'Brien. New York: Vintage, 1942.

Capra, Fritjof. *The Tao of Physics: An Exploration of the Parallels between Modern Physics and Eastern Mysticism*. Boulder, CO: Shambhala, 1975.

Chalmers, Alan F. *What Is This Thing Called Science?* St. Lucia, Aus.: University of Queensland Press, 1976.

Chandler, Daniel. *Semiotics: The Basics*. 2nd ed. Abingdon, UK: Routledge, 2017.

———, and Rod Munday. "Thomas Theorem (Thomas Axiom)." Oxford Reference, 2016. https://www.oxfordreference.com/view/10.1093/acref/9780191800986.001.0001/acref-9780191800986-e-2793.

Childress, Boyd. "Equal Opportunity." Reference for Business, n.d. https://www.referenceforbusiness.com/encyclopedia/Ent-Fac/Equal-Opportunity.html.

Choi, Sungho, and Michael Fara. "Dispositions." In *SEP*, July 26, 2006; revised June 22, 2018. https://plato.stanford.edu/archives/spr2021/entries/dispositions/.

Cicero, Marcus Tullius. *Tusculan Disputations*. Translated by Charles D. Yonge. New York: Harper and Brothers, 1877.

Ciyun, Zhang. *Chinese Idioms and Their Stories*. Beijing: Foreign Languages, 1996.

Clement of Alexandria. *The Stromata*. Translated by William Wilson. Logos Virtual Library, n.d. https://www.logoslibrary.org/clement/stromata/127.html.

Cleveland, Cutler J., ed. *Concise Encyclopedia of the History of Energy*. London: Academic, 2009.

Cohen, S. Marc. "Anaxagoras." Washington University, 2002; updated Sept. 21, 2016. https://faculty.washington.edu/smcohen/320/anaxag.htm.

Coleman, Elizabeth B. "The Disneyland of Cultural Rights to Intellectual Property: Three Concepts in Search of a Relationship." In *Intellectual Property and Traditional*

Cultural Expressions in a Digital Environment, edited by Christof B. Graber and Mira Burri-Nenova, 49–72. Chelenham, UK: Edward Elgar, 2008.

Colestock, Richard O. "How Many Things Are There in the Universe?" Quora, 2018. Reply to online question. https://www.quora.com/How-many-things-are-there-in-the-universe.

Commoner, Barry. "Fallout and Water Pollution: Parallel Cases (Slightly Abridged)." *Scientist and Citizen* 7, no. 2 (Dec. 1964) 2–7. https://doi.org/10.1080/21551278.1964.9958599.

Comte, Auguste. *The Course in Positive Philosophy*. Translated by Harriet Martineau. London: Chapman, 1853.

Conger, George P. "Whitehead Lecture Notes: Seminary in Logic: Logical and Metaphysical Problems." 1927. Manuscripts and Archives, Yale University Library, Yale University, New Haven, CT.

Cooper, Robert L., and Bernard J. Spolsky. *The Influence of Language on Culture and Thought: Essays in Honor of Joshua A. Fishman's Sixty-Fifth Birthday*. Berlin: De Gruyter, 2019.

Copernicus, Nicolaus. *On the Revolutions of Heavenly Spheres*. Translated by Edward Rosen. New York: Prometheus, 2010.

Cowan, George, et al. "Operation Crossroads." Atomic Heritage Foundation, July 1, 2014. https://www.atomicheritage.org/history/operation-crossroads.

Crosland Maurice. "Lavoisier's Theory of Acidity." *Isis* 64 (1973) 306–25.

Croy, Anita. *Galileo Galilei*. Chicago: Raintree. 2021.

Crystal, David. *A Dictionary of Linguistics and Phonetics*. 6th ed. Hoboken, NJ: Blackwell, 2008.

Curran, Jane V., and Christophe Fricker, eds. *Schiller's "On Grace and Dignity" in Its Cultural Context: Essays and a New Translation*. Rochester, NY: Camden House, 2005.

Cyril of Jerusalem. "Catechetical Lecture 15." Translated by Edwin Hamilton Gifford. From *Nicene and Post-Nicene Fathers*, edited by Philip Schaff and Henry Wace, 2nd ser., 7. Buffalo, NY: Christian Literature Publishing Co., 1894. Revised and edited for New Advent by Kevin Knight. https://www.newadvent.org/fathers/310115.htm.

Daniel-Rops, Henri. *History of the Church of Christ*. 8 vols. Boston: Dutton, 1960.

Darwin, Charles. *The Autobiography of Charles Darwin: 1809–1882*. Edited by Nora Barlow. New York: Norton, 1993.

———. *The Origin of Species*. 6th ed. Reprint, Québec City: Feedbooks, n.d.

Dawkins, Richard. *The Evolution of Life*. London: Notting Hill, 1996.

———. "The Improbability of God." *Free Inquiry* 18 (1998) 6–9.

———. *The Selfish Gene*. Oxford, UK: Oxford University Press, 1989.

Dawson, John W. "Gödel and the Limits of Logic." Plus Magazine, June 1, 2006. https://plus.maths.org/content/goumldel-and-limits-logic.

Dee, Sylvia. "The End of the World." Lyrcsia, 1962. https://www.lyricsia.com/lyricid/10597/end-of-the-world.

"Deep Ecology." Lexico, n.d. https://www.lexico.com/definition/deep_ecology.

Defoe, Daniel. *Journal of the Plague Year*. Harlow, UK: Longmans Green, 1895.

DeGraff, Brandon C. "Jane Elliott: 'Racism is a Mental Illness.'" Medium, Sept. 11, 2017. https://medium.com/@brandondegraff/jane-elliott-racism-is-a-mental-illness-2407db6c1962.

Dennett, Daniel C. *Brainstorms: Philosophical Essays on Mind and Psychology*. Cambridge, MA: Massachussetts Institute of Technology Press, 1977.

Denton, William H. "Problem-Solving as a Theory of Learning and Teaching." *High School Journal* 49 (1966) 382–90.
DePillis, Lydia, et al. "America's Addiction to Absurdly Fast Shipping Has a Hidden Cost." *CNN Business*, July 15, 2019. https://depts.washington.edu/sctlctr/news-events/in-the-news/americas-addiction-absurdly-fast-shipping-has-hidden-cost.
Derrida, Jacques. *Limited Inc*. Edited by Gerald Graff, translated by Jeffrey Mehlman and Samuel Weber. Evanston, IL: Northwestern University Press, 1988.
Dervos, Constantine T., and and Vassilou Panayota. "Sulfur Hexafluoride (SF_6): Global Environmental Effects and Toxic Byproduct Formation." *Air and Waste Management Association* 50, no. 1 (2000) 137–41. https://doi.org/10.1080/10473 289.2000.10463996.
Descartes, René. *Discourse on Method and Related Writings*. Translated by John Veitch. New York: Cosimo, 2008.
———. *Meditations on First Philosophy*. Translated by John Veitch. Gearhart, OR: Watchmaker, 2010.
———. *Meditations on First Philosophy with Selections from the Objections and Replies*. Translated by Michael Moriarty. Oxford, UK: Oxford University Press, 2008.
Devitt, Michael. "A Shocking Idea about Meaning." *Revue Internationale de Philosophie* 4 (2001) 471–94.
Dewey, John. "My Pedagogic Creed." *School Journal* 54, no. 1 (1897) 77–80.
Dhahabi, Ilm ar-Rijaal al. *Siyar Alaam an-Nubala*. 25 vols. Beirut: Dar-al-Kotob-al-Ilmiyah, 1990.
D'Holbach, Paul-Henri T. *The System of Nature*. 3 vols. Norderstedt, Germ.: Books on Demand, 2020.
Dilthey, Wilhelm, et al. "Hermeneutics and the Study of History." In *Wilhelm Dilthey: Selected Works*, translated by Rudolf A. Makkreel, 4:229–34. 6 vols. Princeton, NJ: Princeton University Press,1996.
"Discovering the Nucleus: Rutherford's Gold Foil Experiment." Chem Talk, n.d. https://chemistrytalk.org/discovering-the-nucleus-rutherfords-gold-foil-experiment/.
Dobney, Kay, et al. "The Future Cost to the United Kingdom's Railway Network of Heat Related Delays and Buckles Caused by the Predicted Increases in High Summer Temperatures Owing to Climate Change." *Journal of Rail and Rapid Transit*, 224 (2010) 25–34.
Dobrin, Arthur. "Happiness Is How You Are, Not How You Feel." Psychology Today, Jan. 25, 2013. https://www.psychologytoday.com/za/blog/am-i-right/201301/happiness-is-how-you-are-not-how-you-feel.
Domingos, Pedro. *The Master Algorithm: How the Quest for the Ultimate Learning Machine Will Remake Our World*. London: Penguin, 2015.
Douglas, Arthur A., ed. *1955 Quotes of Albert Einstein*. Wiefelstede, Germ.: UB Tech, 2016.
Dryden, John. *The Poetical Works of John Dryden: With the Life of the Author*. 3 vols. Edinburgh: Apollo, 1777.
Duhem, Pierre. *The Aim and Structure of Physical Theory*. 2nd ed. Translated by Philip P. Wiener. Princeton, NJ: Princeton University Press, 1991.
Duignan, Brian. "Postmodernism and Relativism." Britannica, n.d. https://www.britannica.com/topic/postmodernism-philosophy/Postmodernism-and-relativism.
"Durkheim and Social Integration." Social Science Libre Texts, last updated Dec. 15, 2020. https://socialsci.libretexts.org/@go/page/7896.

Durkheim, Emile. *The Division of Labour in Society*. Edited by Steven Lukes. London: Macmillan International Higher Education, 2013.

Dyer, Hugh. "Introducing Green Theory in International Relations." E-International Relations, Jan. 7, 2018. https://www.e-ir.info/2018/01/07/green-theory-in-international-relations/.

"Economics." Lexico, n.d. https://www.lexico.com/definition/economics.

Eddington, Arthur S. *The Nature of the Physical World*. Cambridge: Cambridge University Press, 1948.

Edwards, David Lawrence, and David Lloyd Edwards. *What Anglicans Believe in the Twenty-First Century*. London: Continuum, 2002.

Einstein, Albert. *Essays in Science*. North Chelmsford, MA: Courier Corporation, 2013.

———. *Ideas and Opinions*. 3rd ed. New York: Crown, 1995.

El Ali, Rami A. "On the Intentionality of Emotions." MA thesis, American University of Beirut, 2006.

Elliott, Charlotte. *Selections from the Poems of Charlotte Elliot*. London: Religious Tract Society, 1873.

Epictetus. *The Art of Living: The Classical Manual on Virtue, Happiness, and Effectiveness*. Edited by Sharon Lebell. San Francisco: HarperOne, 2007.

Etzioni, Amitai. *Comparative Analysis of Complex Organizations*. Rev. ed. New York: Simon and Schuster, 1975.

Faleomavaega, Eni F. H., and Jeff Flake. "Commending the People of the Republic of the Marshall Islands." H. R. Res. 692. 109th Congress, 2006. https://www.govinfo.gov/content/pkg/BILLS-109hres692ih/html/BILLS-109hres692ih.htm.

Falguera, José L., et al. "Abstract Objects." In *SEP*, July 19, 2001; revised Aug. 9, 2021. https://plato.stanford.edu/entries/abstract-objects/.

Faulkner, Nicholas, and William L. Hosch, eds. *Numbers and Measurements*. New York: Rosen, 1905.

Ferguson, Kitty. *The Fire in the Equations: Science Religion and Search for God*. Philadelphia: Templeton, 2004.

Feyerabend, Paul K. *Against Method: Outline of an Anarchistic Theory of Knowledge*. London: Verso, 1978.

———. *Killing Time: The Autobiography of Paul Feyerabend*. Chicago: University of Chicago Press, 1996.

Feynman, Richard P. *The Character of Physical Law*. Cambridge, MA: Massachussetts Institute of Technology, 2017.

———. "Feynman: Take the World From Another Point of View." YouTube, May 11 2008. https://www.youtube.com/watch?v=PsgBtOVzHKI.

———, et al. *The Feynman Lectures on Physics*. 3 vols. Boston: Addison-Wesley, 1977.

Fillingham, Lydia A. *Foucault for Beginners*. Danbury, CT: Writers and Readers, 1993.

Fine, Arthur. "Einstein, Albert." In *CDP*, 219–20.

Fletcher, Ronald. "Plural Society." In *FDMT*, 655.

Foot, Philippa. *Natural Goodness*. Oxford, UK: Clarendon, 2003.

"Forbes Quotes: Thoughts on the Business of Life." *Forbes*, 2015. https://www.forbes.com/quotes/8858/.

Fortuna, Luigi, et al. *Five Hundred Years after Leonardo DaVinci Machines: Towards Innovation and Control*. Singapore: World Scientific, 2020.

Fox, Melodie J. "Prototype Theory: An Alternative Concept Theory for Categorizing Sex and Gender?" *Proceedings from North American Symposium on Knowledge Organization* 3 (2011) 151–59.

Fox, Nick J., and Pam Alldred. *Sociology and the New Materialism: Theory, Research, Action*. Los Angeles: SAGE, 2016.
Francis, Pope. "*Laudato Si'*: On Care for our Common Home." Vatican, May 24, 2015. https://www.vatican.va/content/francesco/en/encyclicals/documents/papa-francesco_20150524_enciclica-laudato-si.html.
Frank, Phillipp. *Einstein: His Life and Times*. New York: Knopf, 1947.
Frankel, Charles. *The Case for Modern Man*. Boston: Beacon, 1956.
Free, Ann Cottrell. *Animals, Nature and Albert Schweitzer*. Graz, Aus.: Albert Schweitzer Center, 1988.
Frege, Gottlob. *Philosophy of Logic*. Edited by Michael Beaney and Erich H. Reck. Abingdon-on-Thames, UK: Taylor and Francis, 2005.
Gardner, Andy. "Price's Equation Made Clear." *Philosophical Transactions of the Royal Society* 355 (2019) 361.
Gassendi, Pierre. "Peace of Mind." In *Three Discourses of Happiness, Virtue, and Liberty*. Translated by François Bernier. London: Awnsham and Churchill, 1699.
Gelernter, David. *The Muse in the Machine: Computerizing the Poetry of Human Thought*. New York: Simon and Schuster, 2010.
Gessen, Masha. "Judith Butler Wants Us to Reshape Our Rage." *New Yorker*, Feb. 9, 2020. https://www.newyorker.com/culture/the-new-yorker-interview/judith-butler-wants-us-to-reshape-our-rage.
Geyrhalter, Fabian, and David Glaze. *Bigger Than This: How to Turn Any Venture into an Admired Brand*. Los Angeles: Brandtro, 2018.
Girifalco, Louis. *The Universal Force: Gravity, the Creator of Worlds*. Oxford, UK: Oxford University Press, 2007.
Giubilini, Alberto. "Conscience." In *SEP*, Mar. 14, 2016; revised Feb. 11, 2021. https://plato.stanford.edu/entries/conscience/.
Glazer, Nathan. "Why Ethnicity?" *Commentary*, Oct. 1974. https://www.commentary.org/articles/nathan-glazer-2/why-ethnicity/.
Golomb, Jacob. "Sartre's Early Phenomenology of Authenticity in Relation to Husserl." *Analecta Husserliana* 80 (2002) 335–42.
Goodwin, Samuel. *Just How Dumb Do You Think I Am?* Maitland, FL: Xulon, 2008.
Graham, Daniel W. "Heraclitus." In *SEP*, Feb. 8, 2007; revised Sept. 3, 2019. https://plato.stanford.edu/archives/sum2021/entries/heraclitus/.
Grayling, Anthony C. "Individuation." In *FDMT*, 416.
Grimshaw, Anna. "Symbol." In *FDMT*, 836.
Hahn, Roger. *Pierre Simon Laplace, 1749–1827: A Determined Scientist*. Cambridge, MA: Harvard University Press, 2005.
Halcomb, James, et al. *Courageous Leaders Transforming Their World*. Seattle: YWAM, 2000.
Halliday, Michael A. K., and Christian M. I. M. Matthiessen. *Halliday's Introduction to Functional Grammar*. Abingdon, UK: Routledge, 2013.
Hanasz, Waldemar. "The Common Good in Machiavelli." *History of Political Thought* 31 (2010) 57–85.
Harris, Eirik Lang. "Legalism." Oxford Bibliographies, June 8, 2017; last modified June 27, 2018. https://www.oxfordbibliographies.com/view/document/obo-9780199920082/obo-9780199920082-0079.xml.
Harris, James C. *Developmental Neuropsychiatry*. 2 vols. Oxford, UK: Oxford University Press, 1998.
Hart, Kevin. *Postmodernism: A Beginner's Guide*. Oxford, UK: One World, 2004.

Hawking, Stephen. *A Brief History of Time*. New York: Bantam, 1998.
———. "Stephen Hawking Interview: Last Week Tonight with John Oliver (HBO)." YouTube, June 15, 2014. https://www.youtube.com/watch?v=T8y5EXFMD4s.
———. "Stephen Hawking Quotes: Most Memorable." Foreign Policy, May 21, 2019. https://foreignpolicyi.org/stephen-hawking-quotes-most-memorable/
———, and Leonard Mlodinow. *The Grand Design*. New York: Random House, 2010.
Hayakawa, Samuel I. *Language in Thought and Action*. London: Allen and Unwin, 1977.
Hegel, Georg W. F. *Lectures on the History of Philosophy*. 3 vols. Translated by Elizabeth S. Haldane and Frances H. Simson. Lincoln: University of Nebraska Press, 1995.
———. *The Positivity of the Christian Religion*. Marxists Internet Archive, 1795. https://www.marxists.org/reference/archive/hegel/works/pc/ch01.htm.
Heidegger, Martin. *The Question Concerning Technology, and Other Essays*. New York: Garland, 1977.
Hemingway, Ernest. *A Farewell to Arms*. New York: Scribner, 2014.
Hepburn, Brian, and Hanne Andersen. "Scientific Method." In *SEP*, Nov. 13, 2015; revised June 1, 2021. https://plato.stanford.edu/entries/scientific-method/.
Heraclitus. *Fragments*. Translated by Brooks Haxton. London: Penguin Classics, 2003.
Hermans, Cor. *Interbellum Literature: Writing in a Season of Nihilism*. Leiden, Neth.: Brill, 2017.
Heshmat, Shahram. "Overcoming Negative Self-Thinking." Psychology Today, Mar. 18, 2015. https://www.psychologytoday.com/gb/blog/science-choice/201503/overcoming-negative-self-thinking-0.
Hesse, Hermann. *Journey to the East*. Translated by Hilda Rosner. Torrington, CT: Martino, 2011.
———. *Siddharta*. Smyrna, DE: Prestwick, 2005.
Hesse, Mary B. *Forces and Fields: The Concept of Action at a Distance in the History of Physics*. North Chelmsford, MA: Courier Corporation, 2005.
Hiton, Lisa. *The Theory of Relativity*. Buffalo, NY: Cavendish Square, 2016.
Hofstadter, Douglas R. *Gödel, Escher, Bach: An Eternal Golden Braid*. New York: Basic, 1994.
———, and Emmanuel Sander. *Surfaces and Essences: Analogy as the Fuel and Fire of Thinking*. New York: Basic, 2013.
Hogarth, William. *The Analysis of Beauty: The Infinite Complexities of Beauty in Six Principles*. Leicester-Fields, UK: Hogarth, 1753. https://arthistoryproject.com/artists/william-hogarth/the-analysis-of-beauty-chapter-5-intricacy/.
Holder, Rodney D. *Nothing but Atoms and Molecules? Probing the Limits of Science*. Tunbridge Wells, UK: Monarch, 1993.
Hook, Sidney. *Sidney Hook on Pragmatism, Democracy, and Freedom: The Essential Essays*. Edited by Robert B. Talisse and Robert Tempio. New York: Prometheus, 2002.
Horgan, John. "David Bohm, Quantum Mechanics and Enlightenment." *Scientific American*, July 23, 2018. https://blogs.scientificamerican.com/cross-check/david-bohm-quantum-mechanics-and-enlightenment/.
———. "From Complexity to Perplexity." *Scientific American* 272 (1995) 104–9.
Hoyningen-Huene, Paul. *Systematicity: The Nature of Science*. New York: Oxford University Press, 2013.
Hume, David. *An Inquiry Concerning Human Understanding*. North Chelmsford, MA: Courier, 2004.
Huxley, Thomas H. *On the Physical Basis of Life*. New York: Van Nostrand, 1871.

Jacquette, Dale. *The Philosophy of Schopenhauer*. Continental European Philosophy 6. Abingdon, UK: Routledge, 2015.

Jakobson, Roman. "On Linguistic Aspects of Translation." In *On Translation*, edited by Reuben A. Brower, 232–39. Cambridge, MA: Harvard University Press, 2013.

———. *To Honor Roman Jakobson: Essays on the Occasion of His Seventieth Birthday, 11 October 1966*. 3 vols. The Hague: Mouton, 1967.

James, William. *The Will to Believe*. Norderstedt, Germ.: Books on Demand, 2018.

Janet, Pierre. *L'Automatisme Psychologique: Essai de Psychologie Expérimentale sur les Formes Inférieures de l'Activité Humaine*. [*Of Psychological Automatism*]. 2 vols. Paris: Alcan, 1889.

Jaspers, Karl. *Way to Wisdom: An Introduction to Philosophy*. Translated by Ralph Manheim. New Haven, CT: Yale University Press, 2003.

Jefferson, Thomas. *The Writings of Thomas Jefferson*. Washington, DC: Taylor and Maury, 1854.

Jenkins, David. "The Lockean Constitution: Separation of Powers and the Limits of Prerogative." *McGill Law Journal* 56 (2011) 489–773.

Johnson, Robert, and Adam Cureton. "Kant's Moral Philosophy." In *SEP*, Feb. 23, 2004; revised July 7, 2016. https://plato.stanford.edu/archives/spr2021/entries/kant-moral/.

Jones, Karen S. "Index Term Weighting." *Information Storage and Retrieval* 11 (1973) 619–33.

Kamlah, Wilhelm, and Paul Lorenzen. *Logical Propaedeutic: Pre-School of Reasonable Discourse*. Lanham, MD: University Press of America, 1984.

Kant, Immanuel. *Critique of Pure Reason*. Translated by John M. D. Meiklejohn. London: Everyman's Library, 1984.

———. *Groundwork for the Metaphysics of Morals*. Translated by Christopher Bennett et al. Oxford, UK: Oxford University Press, 2019.

———. *Religion within the Limits of Bare Reason*. Translated by Werner S. Pluhar. Indianapolis: Hackett, 2009.

Kaplan, Aryeh M. E. *The Aryeh Kaplan Reader: The Gift He Left Behind: Collected Essays on Jewish Themes from the Noted Writer and Thinker*. Rahway, NJ: Mesorah, 1986.

Kates, Joshua. *A New Philosophy of Discourse: Language Unbound*. London: Bloomsbury, 2021.

Kauffman, Kohler. "Immortality of the Soul." In *Jewish Encyclopedia*, edited by Joseph Jacobs, 6:564–67. New York: Funk and Wagnalls, 1906.

Kauffman, Stuart. "Why Is the Universe Complex? Broken Symmetries, Information, Energy, Work." *NPR*, June 30, 2010. https://www.npr.org/sections/13.7/2010/06/30/128212122/why-is-the-universe-complex-broken-symmetries-information-energy-work.

Kennedy, John F. "The President and the Press: Address before the American Newspaper Publishers Association." JFK Library, Apr. 7, 1961. https://www.jfklibrary.org/archives/other-resources/john-f-kennedy-speeches/american-newspaper-publishers-association-19610427.

Kenyon, Karl W., and Eugene Kridler. "Laysan Albatrosses Swallow Indigestible Matter." *Auk* 86 (1969) 339–43.

Kierkegaard, Søren. *Either/Or*. Translated by Howard V. Hong and Edna H. Hong. Princeton, NJ: Princeton University Press, 1988.

———. *Works of Love*. Translated by David F. Swenson and Lillian M. Swenson. Princeton, NJ: Princeton Universtity Press, 1949.

Bibliography

King, Martin Luther, Jr. *Strength to Love*. New York: Penguin Random House, 2019.
Kipling, Rudyard. *Rewards and Fairies*. Oxford, UK: Palala, 2015.
Koestler-Grack, Rachel A. *Leonardo Da Vinci: Artist, Inventor, and Renaissance Man*. New York: Infobase, 2005.
Korten, David C. *When Corporations Rule the World*. 2nd ed. Boulder, CO: Kumarian, 2001.
Korzybski, Alfred. *Science and Sanity: An Introduction to Non-Aristotelian Systems and General Semantics*. Forest Hills, NY: Institute of General Semantics, 1958.
Kraut, Richard. "Aristotle's Ethics." In *SEP*, May 1, 2001; revised June 15, 2018. https://plato.stanford.edu/archives/sum2018/entries/aristotle-ethics/
Kuhn, Thomas. *The Structure of Scientific Revolutions*. 3rd ed. Chicago: University of Chicago Press, 1996.
Küng, Hans. *On Being a Christian*. Translated by Edward Quinn. New York: Doubleday, 1974.
Kusama, quoted in Guggenheim Museum, "Forget yourself," Twitter, Mar. 22, 2017, 12:05 p.m., https://twitter.com/guggenheim/status/844580729170378752.
Kuznicki, Jason. "A Kantian Case for Libertarianism." Libertarianism, Jan 3, 2017. https://www.libertarianism.org/publications/essays/kantian-case-libertarianism.
LaBoy, Felicia H. *Table Matters: The Sacraments, Evangelism, and Social Justice*. Eugene, OR: Wipf and Stock, 2017.
Laertius, Diogenes. *Complete Works of Diogenes Laertius*. Translated by Robert D. Hicks. Hastings, UK: Delphi Classics, 2015.
Lambert, John, ed. *Vespers-Book: Containing the Complete Order for Vespers for the Entire Year, According to the Roman Breviary*. London: Burns and Oates, 1850.
Lamey, Andy. "Ecosystems as Spontaneous Orders." *Critical Review: Journal of Politics and Society* 27 (2015) 64–88.
Laozi (Lao Tzu). *Almost Paradise: New and Selected Poems and Translations*. Translated by Sam Hamill. Boulder, CO: Shambhala, 2005.
———. *Tao te Ching: A New Translation*. Translated by Sam Hamill. Boulder, CO: Shambhala, 2005.
Lapchak, Paul A., and Paul D. Boitano. "Reflections on Neuroprotection Research and the Path toward Clinical Success." In *Neuroprotective Therapy for Stroke and Ischemic Disease*, edited by Paul A. Lapchak and Paul D. Boitano, 3–72. Springer Series in Transitional Stroke Research. Berlin: Springer, 2017.
Lavoisier, Antoine. *Elements of Chemistry*. Translated by Robert Kerr. Edinburgh: Creech, 1790. https://web.lemoyne.edu/~giunta/ea/lavprefann.html.
Leary, Timothy. *The Politics of Ecstasy*. Berkeley, CA: Ronin, 1968.
Leech, Geoffrey N. *Semantics: The Study of Meaning*. London: Penguin, 1974.
Levenson, Eric. "How Minneapolis Police First Described the Murder of George Floyd, and What We Know Now." *CNN*, Apr. 21, 2021. https://edition.cnn.com/2021/04/21/us/minneapolis-police-george-floyd-death/index.html.
Levitt, Michael. "There Is an Environmental Impact Each Time You Hit 'Buy Now.' Here's an Alternative." *NPR*, Dec. 5, 2021. https://www.npr.org/2021/12/05/1060989079/looking-for-holiday-shopping-sales-theres-a-low-impact-option.
Lewis, Hywel D. *Persons and Life after Death*. London: Palgrave Macmillan, 1977.
Lindberg, David C. *The Beginnings of Western Science*. Chicago: University of Chicago Press, 1992.
Löbner, Sebastian. *Understanding Semantics*. Abingdon, UK: Routledge, 2013.

Locke, John. *An Essay Concerning Human Understanding*. Edited by Peter H. Nidditch. Oxford, UK: Clarendon, 1975.

———. *The Educational Writings of John Locke*. Edited by John W. Adamson. Cambridge: Cambridge University Press, 1922.

———. *Two Treatises of Government*. Cambridge: Cambridge University Press, 1967.

Losee, John. *A Historical Introduction to the Philosophy of Science*. Oxford, UK: Oxford University Press, 2001.

Losonsky, Michael. *Linguistic Turns in Modern Philosophy*. Cambridge: Cambridge University Press, 2006.

Lucas, John R. "The Implications of Gödel's Theorem." *Ethics and Politics* 5 (2003) 1. https://www2.units.it/etica/2003_1/14_monographica.htm.

Luntley, Michael. *Reason, Truth and Self: The Postmodern Reconditioned*. New York: Routledge, 1995.

Lynch, William F., ed. "Ugliness." In *New Catholic Encyclopedia*, 1:270–71. Detroit: Thomson Gale, 2003.

Lyotard, Jean-François. *Le Différend*. Minneapolis: University of Minnesota Press, 1988.

———. *The Postmodern Condition: A Report on Knowledge*. Translated by Geoff Bennington and Brian Massumi. Minneapolis: University of Minnesota, 1984.

Lyth, David. "How Einstein's General Theory of Relativity Killed Off Common Sense Physics." Conversation, Nov. 24, 2015. https://theconversation.com/how-einsteins-general-theory-of-relativity-killed-off-common-sense-physics-50042.

MacBride, Fraser, "Relations." In SEP, Feb. 9, 2016; revised Oct. 28, 2020.

Mach, Ernst. *The Science of Mechanics: A Critical and Historical Account of Its Development*. Chicago: Open Court, 1893.

Madison, James. "To the People of New York." In *Federalist Papers* 51 (1788).

"Main Concepts of Confucianism." Lander University, n.d. https://philosophy.lander.edu/oriental/main.html.

Malin, Shilon. *Nature Loves to Hide: Quantum Physics and the Nature of Reality, a Western Perspective*. Rev. ed. Singapore: World Scientific, 2012.

Mannan, Abdul. "Science and Subjectivity: Understanding Objectivity of Scientific Knowledge." *Philosophy and Progress* 59–60 (2016) 1607–2278.

Manning, Russell R. *The Cambridge Companion to Paul Tillich*. Cambridge: Cambridge University Press, 2009.

Marcuse, Herbert. "The Individual in the Great Society. Part II." *Alternatives Journal*, 6 (1966) 29–32.

Margolis, Eric, and Stephen Laurence. "Concepts." In SEP, Nov. 7, 2005; revised June 17, 2019. https://plato.stanford.edu/archives/spr2021/entries/concepts/.

Marx, Karl. *Capital: A Critique of Political Economy*. Edited by Frederick Engels. Translated by Samuel Moore and Edward Aveling. 3 vols. Marxists Internet Archive, 1867, 1887. https://www.marxists.org/archive/marx/works/1867-c1/.

Matthews, Peter H. *A Short History of Structural Linguistics*. Cambridge: Cambridge University Press, 2001.

———, ed. *The Concise Oxford Dictionary of Linguistics*. 3rd ed. Oxford, UK: Oxford University Press, 2014.

Mautner, Thomas, ed. *The Penguin Dictionary of Philosophy*. London: Penguin Reference, 2000.

Mayo-Wilson, Conor. "Plato's Theory of Recollection." University of Washington, Jan. 31, 2017. http://faculty.washington.edu/conormw/Teaching/Files/PhilMath/Winter_2017/Lectures/7_Lecture_373_Win_2017.pdf.

McAleer, Sean. *Plato's Republic: An Introduction*. Cambridge, UK: Open Book, 2020.
McDowell, John. "Aesthetic Value, Objectivity, and the Fabric of the World." In *Pleasure, Preference, and Value: Studies in Philosophical Aesthetics*, edited by Eva Schaper, 1–16. Cambridge: Cambridge University Press, 1983.
McLuhan, Marshall, and W. Terrence Gordon. *Understanding Media: The Extensions of Man*. Hamburg: Gingko, 2003.
Medema, Steven G. *The Hesitant Hand: Taming Self-Interest in the History of Economic Ideas*. Princeton, NJ: Princeton University Press, 2009.
Melanby, Kenneth. "Environment." In *FDMT*, 275.
Menahem, Ari Ben. *Historical Encyclopedia of Natural and Mathematical Sciences*. 6 vols. Berlin: Springer, 2009.
Midgley, Mary. "Can Science Save Its Soul?" New Scientist, July 31, 1992. https://www.newscientist.com/article/mg13518324-300-can-science-save-its-soul-some-scientists-have-begun-to-talk-confidently-about-understanding-god-and-creation-they-are-crediting-science-with-power-it-doesnt-possess/.
Miller, Nicola. "Kirkpatrick Theory." In *FDMT*, 456.
MIT Mechanical Engineering. "A Life-Saving Discovery at MIT MechE." YouTube, Dec. 29, 2015. https://www.youtube.com/watch?v=36_pUKGOC-4.
Mole, Christopher. "Attention." In *SEP*, Sept. 8, 2009; revised Oct. 26, 2021. https://plato.stanford.edu/entries/attention/.
Mommaers, Paul. *The Riddle of Christian Mystical Experience: The Role of the Humanity of Jesus*. Leuven, Belg.: Peeters, 2003.
Morgan, Michael. "Galileo's Pendulum Experiments." Rice University, last revised Apr. 17, 1995. http://galileo.rice.edu/lib/student_work/experiment95/galileo_pendulum.html.
Moyn, Samuel. "Why Do Americans Have So Few Rights?" *New Republic*, Mar 9, 2021. https://newrepublic.com/article/161561/americans-rights-jamal-greene-book-review.
Mulvaney, Robert J., ed. *Classic Philosophical Questions*. 14th ed. Columbia: University of South Carolina Press, 2011.
Munitz, Milton. *Contemporary Analytic Philosophy*. London: Pearson, 1981.
Nagel, Thomas. *What Does It All Mean? A Very Short Introduction to Philosophy*. Oxford, UK: Oxford University Press, 1987.
Needleman, Jacob, and David Appelbaum. *Real Philosophy: An Anthology of the Universal Search for Meaning*. Fayetteville, AR: Arkana, 1990.
Newton, Isaac. *The Principia: Mathematical Principles of Natural Philosophy*. Oakland, CA: University of California Press, 1999.
Ng, Ruth. "What Is an Object? A Philosophical Exploration." Medium, Jan. 23, 2018. https://medium.com/@ruth.ym.ng/what-is-an-object-a-philosophical-exploration-85cd84d3f878.
Niebuhr, Reinhold. *Faith and History: A Comparison of Christian and Modern Views of History*. New York: Scribners and Sons, 1949.
Nietzsche, Friedrich W. *Twilight of the Idols and The Anti-Christ*. Translated by Reginald J. Hollingdale. Harmondsworth, UK: Penguin, 1968.
———. *The Will to Power*. Edited by Walter Kaufmann. Translated by Walter Kaufmann and Reginald J. Hollingdale. New York: Vintage, 1968.
O'Connor, John J., and Edmund F. Robertson. "Georg Simon Ohm." MacTutor, last updated Jan. 2000. https://mathshistory.st-andrews.ac.uk/Biographies/Ohm/.

O'Leary, Zina. *The Social Science Jargon Buster: The Key Terms You Need to Know.* Los Angeles: SAGE, 2007.

Opp, Karl-Dieter. "What Is Is Always Becoming What Ought to Be: How Political Action Generates a Participation Norm." *European Sociological Review* 20 (2004) 13–29.

Oreskes, Naomi, et al. "Verification, Validation, and Confirmation of Numerical Models in the Earth Sciences." *Science, New Series* 263 (1994) 641–46.

Orzel, Chad. "What Are the Limits of Physics?" *Forbes,* June 15, 2015. https://www.forbes.com/sites/chadorzel/2015/06/15/what-are-the-limits-of-physics/?sh=533db01cc585.

Osborne, Richard. *Philosophy for Beginners.* Danbury, CT: For Beginners, 2007.

Osgood, Charles. "Where Do Sentences Come From?" In *Semantics: An Interdisciplinary Reader in Philosophy, Linguistics and Psychology,* edited by Danny D. Steinberg and Leon A. Jakobovits, 497–529. Cambridge: Cambridge University Press Archive, 1970.

"Pablo Picasso 1881–1973." Oxford Reference, n.d. https://www.oxfordreference.com/view/10.1093/acref/9780191843730.001.0001/q-oro-ed5-00008311.

Paech, Nico. "The Destructive Dream of Progress." Degrowth, Jan 19, 2017. https://www.degrowth.info/en/blog/the-destructive-dream-of-progress.

Parfit, Derek. *On What Matters.* 3 vols. Oxford, UK: Oxford University Press, 2013.

Pascal, Blaise. *The Thoughts of Blaise Pascal.* Translated by Charles K. Paul. Scotts Valley, CA: CreateSpace, 2015.

Paton, Alan. *Diepkloof: Reflections of Diepkloof Reformatory.* Edited by Clyde Broster. Cape Town: New Africa Books, 1986.

Pecorino, Philip A. *Philosophy of Religion: Online Textbook.* Queensborough Community College, 2001. https://www.qcc.cuny.edu/socialsciences/ppecorino/phil_of_religion_text/default.htm.

Peek, Ella. "Ethical Criticism of Art." Internet Encyclopedia of Philosophy, n.d. https://iep.utm.edu/art-eth/.

Peirce, Charles S. *Collected Papers of Charles Sanders Peirce.* 8 vols. Cambridge, MA: Harvard University Press, 1974.

Peterson, Jordan B. "An Atheist in the Realm of Myth." Interview with Stephen Fry. YouTube, May 17, 2021; orig. Jordan B. Peterson podcast S4:E22, Mar. 28, 2021. https://www.youtube.com/watch?v=fFFSKedy9f4.

Pettazzoni, Raffaele. *Essays on the History of Religions.* Translated by Herbert. J. Rose. Studies in the History of Religions 1. Leiden, Neth.: Brill, 1954.

Philp, Mark. "Thomas Paine." In *SEP,* July 18, 2013; revised Aug. 16, 2021. https://plato.stanford.edu/entries/paine/.

Pichler, Alois, and Simo Säätelä. *Wittgenstein: The Philosopher and His Works.* 4 vols. Berlin: Ontos, 2006.

Pinker, Steven. "How to Avoid Falling for the 'Gambler's Fallacy.'" *BBC,* Dec. 16, 2021. https://www.bbc.co.uk/ideas/videos/how-to-avoid-falling-for-the-gamblers-fallacy/pobb6sj8.

Pitt, Joseph C. *Theories of Explanation.* Oxford, UK: Oxford University Press. 1988.

Plato. *The Apology of Socrates.* Edited by E. E. Garvin. Translated by H. N. Fowler. University of Alberta, 1913, 2013. https://sites.ualberta.ca/~egarvin/assets/plato-apology.pdf.

———. *The Dialogues of Plato.* Translated by Benjamin Jowett. 5 vols. Oxford, UK: Clarendon, 1892.

———. *The Republic of Plato*. 2nd ed. Translated by Allan Bloom. New York: Basic, 1991.

Polanyi, Michael. *The Study of Man: The Lindsay Memorial Lectures*. Abingdon, UK: Routledge Revivals, 2013.

Poli, Roberto, et al. *The Philosophy of Nicolai Hartmann*. Berlin: De Gruyter, 2011.

Porst, Jana, and Avi Sadeh. "Imagination-Based Interventions with Children." In *The Oxford Handbook of the Development of Imagination*, edited by Marjorie Taylor, 529–38. Oxford, UK: Oxford University Press, 2013.

Porth, Eric, et al. "Functionalism." University of Alabama, n.d. https://anthropology.ua.edu/theory/functionalism/.

Putnam, Ruth A. "James, William." In *Encyclopedia of Ethics*, edited by Lawrence C. Becker and Charlotte B. Becker, 899–901. Abingdon, UK, Routledge, 2013.

Quine, Willard O., and Joseph S. Ullian. *The Web of Belief*. 2nd ed. New York: McGraw Hill, 1978.

Quinton, Anthony. "Concept." In *FDMT*, 159.

Ratcliffe, Susan, ed. *The Concise Oxford Dictionary of Quotations*. New York: Oxford University Press, 2011.

Raturi, Prerna. "What You Seek Is Seeking You: Rumi for Entrepreneurs." *Entrepreneur*, Jan. 18, 2016. https://www.entrepreneur.com/article/269709.

Raypole, Crystal. "People Can Change, but That Doesn't Mean They Will." *Healthline*, July 31, 2020. https://www.healthline.com/health/do-people-change.

Read, Piers P. *The Dreyfus Affair*. London: Bloomsbury, 2013.

Redaelli, Simone. "A Guide to Using the Scientific Method in Everyday Life." PLOS SciComm, Aug. 4, 2020. https://scicomm.plos.org/2020/08/04/a-guide-to-using-the-scientific-method-in-everyday-life/.

Redding, Paul. "Georg Wilhelm Friedrich Hegel." In *SEP*, Feb. 13, 1997; revised Oct. 20, 1997. https://stanford.library.sydney.edu.au/archives/sum2001/entries/hegel/.

Rettler, Bradley, and Andrew M. Bailey. "Object." In *SEP*, Oct. 26, 2017. https://plato.stanford.edu/entries/object/.

Reuell, Peter. "Cities' Wealth Gap Is Growing, Too." *Harvard Gazette*, May 2, 2019. https://news.harvard.edu/gazette/story/2019/05/study-finds-gap-between-rich-and-poor-growing-regionally-too/.

Rico, Luis. "Spanish Heritage in Mathematics and Mathematics Education." *The Proceedings of the 12th International Congress on Mathematical Education* 12 (2015) 331–41.

Roberts, Peter. *George Costakis: A Russian Life in Art*. Montreal: McGill-Queen's University Press, 1994.

Robeyns, Ingrid, and Morten F. Byskov. "The Capability Approach." In *SEP*, Apr. 14, 2011; revised Dec. 10, 2020. https://plato.stanford.edu/entries/capability-approach/

Rolston, Holmes, III. "Nature for Real: Is Nature a Social Construct?" In *The Philosophy of the Environment*, edited by Timothy D. J. Chappell, 38–64. Edinburgh: University of Edinburgh Press, 1997.

Rosensohn, William L. *The Phenomenology of Charles S. Peirce: From the Doctrine of Categories to Phaneroscopy*. Amsterdam: Benjamins, 1974.

Rousseau, Jean Jacques. *Emile*. Translated by Barbara Foxley. Mineola, NY: Dover, 2013.

Rowe, William L. "Tillich's Theory of Signs and Symbols." *Monist* 4 (1966) 593–610.

Ruskin, John. *The Genius of John Ruskin: Selections from His Writings*. Edited by John D. Rosenberg. Charlottesville: University of Virginia Press, 1997.

Russell, Bertrand. *Our Knowledge of the External World as a Field for Scientific Method in Philosophy*. London: Allen and Unwin, 1914.

———. *The Problems of Philosophy*. Oxford, UK: Oxford University Press, 1980.

Rutherford, Ernest. "Rutherford's Incredible Event." Pomona College, Oct. 27, 2010. From Rutherford, "The Development of the Theory of Atomic Structure," lecture, 1936. http://carneades.pomona.edu/2010-PoP/1027-hdo.pdf.

Saint-Simon, Claude-Henri de. *The Political Thought of Saint-Simon*. Edited by Ghita Ionescu. Oxford, UK: Oxford University Press, 1976.

Saito, Yuriko. "Everyday Aesthetics." *Philosophy and Literature* 49 (2009) 2.

Sánchez, Francisco. *Nothing Is Known*. Cambridge: Cambridge University Press, 1988.

Sartre, Jean-Paul. *Jean-Paul Sartre: Basic Writings*. Abingdon:, UK Routledge, 2002.

Saussure, Ferdinand de. *Course in General Linguistics*. Edited by Perry Meisel and Haun Saussy. Translated by Wade Baskin. New York: Columbia University Press, 2011.

Sautet, Marc. *Nietzsche for Beginners*. Illustrated by Patrick Boussignac. Danbury, CT: For Beginners, 2007. Kindle edition.

Savile, Anthony. *Leibniz and the Monadology*. Routledge Philosophy GuideBook. Abingdon, UK: Routledge, 2002.

Sayer, R. Andrew. *Method in Social Science: A Realist Approach*. Hove, UK: Psychology, 1992.

Scarborough, Thomas O. "CCO Metal Detector." Elektor, July 2005; last updated Dec. 30, 2017. https://www.elektormagazine.com/magazine/elektor-200507/18074.

———. "Christian Transformational Leadership: A Deconstructionist Critique." MTh thesis, South African Theological Seminary, 2009.

———. "Metaphysical Notes." Philosopher, 2015. On-line publication. https://www.dropbox.com/s/rqwb8zy5op99coj/metaphysicsal-notes.pdf.

———. "Revisiting Aristotle's Noun." *Philosopher* 101, no. 1 (Spring 2013) n.p. http://www.the-philosopher.co.uk/2013/05/revisiting-aristotles-noun.html.

Schaeffer, Francis A. *The God Who Is There*. Downers Grove, IL: InterVarsity, 2020.

Schlick, Moritz. *Moritz Schlick Philosophy Papers: Volume 1 (1909–22)*. Edited by Henk L. Mulder and Barbara F. B. van de Velde-Schlick. Berlin: Springer, 1978.

Schopenhauer, Arthur. *The World as Will and Representation*. Translated by Eric F. J. Payne. Toronto: Aegitas, 2016.

Schrödinger, Erwin. *What Is Life?: With Mind and Matter and Autobiographical Sketches*. Cambridge: Cambridge University Press, 2012

Schulz, Ortrun. *Schopenhauer's Critique of Hope*. Norderstedt, Germ.: Books on Demand, 2014.

Scruton, Roger. *Kant: A Very Short Introduction*. Oxford, UK: Oxford University Press, 2001.

Seachris, Joshua W., ed. *Exploring the Meaning of Life: An Anthology and Guide*. Hoboken, NJ: Wiley and Sons, 2012.

Searle, John R. *Speech Acts*. Cambridge: Cambridge University Press, 1969.

Seidman, Steven. *The Postmodern Turn: New Perspectives on Social Theory*. Cambridge: Cambridge University Press, 1994.

Seiler, Robert M. "Human Communication in the Critical Theory Tradition." University of Calgary, 2006. https://people.ucalgary.ca/~rseiler/critical.htm.

Seretan, Violeta. *Syntax-Based Collocation Extraction*. New York: Springer Science and Business, 2011.

Shaw, John F. *New Dictionary of Quotations from the Greek, Latin, and Modern Languages*. Philadelphia: Lippincott, 1859.

Shirazi, Sayyid Abdul Husayn Dastghaib. "The Fortieth Greater Sin: Persistence in Minor Sins." Translated by Sayyid Athar Husayn S. H. Rizvi. Al-Islam, 2004. https://www.al-islam.org/greater-sins-volume-3-sayyid-abdul-husayn-dastghaib-shirazi/fortieth-greater-sin-persistence-minor.

Shukman, David. "Hawking: Humans at Risk of Lethal 'Own Goal.'" *BBC*, Jan. 19, 2016. https://www.bbc.com/news/science-environment-35344664.

Sia, Paolo di. "Describing the Concept of Infinite among Art, Literature, Philosophy and Science: A Pedagogical-Didactic Overview." *Journal of Education Culture and Society* 5 (2014) 9–19.

Siegel, Ethan. "No, the Universe Is Not Purely Mathematical in Nature." *Forbes*, May 20, 2020. https://www.forbes.com/sites/startswithabang/2020/05/20/no-the-universe-is-probably-not-mathematical-in-nature/?sh=5a60bbad1165.

Simanek, Donald, compiler. "Science Quotes." Lock Haven University, last revised June 10, 2001. https://www.lockhaven.edu/~dsimanek/sciquote.htm.

Simon, Yves R. *Philosophy of Democratic Government*. Notre Dame, IN: Notre Dame Press, 1993. https://maritain.nd.edu/jmc/etext/pdg.htm.

Sinclair, John M. "Subjective." In *The Collins English Dictionary*. Glasgow: Collins, 1998.

Sloman, Aaron. *The Computer Revolution in Philosophy: Philosophy, Science, and Models of Mind*. Hassocks, UK: Harvester, 1978. Online edition last updated Jan. 2020. https://www.cs.bham.ac.uk/research/projects/cogaff/crp/.

Smith, Daniel. *How to Think Like a Philosopher*. London: O'Mara, 2021. Kindle edition.

Soanes, Catherine. "Must, Should, or Ought To?" Calameo, Mar. 3, 2014. https://en.calameo.com/read/0053777387bf2aa578dcc.

Sobosan, Jeffrey G. *Romancing the Universe: Theology, Cosmology, and Science*. Grand Rapids: William B. Eerdmans, 1999.

Spencer, Lloyd, and Andrzej Krauze. *Introducing Hegel*. Cambridge, UK: Icon, 1999.

Spinoza, Benedictus de. *The Chief Works of Benedict de Spinoza*. Translated by Robert H. M. Elwes. 2 vols. 2nd ed. London: George Bell and Sons, 1887.

Stang, Nicholas F. "Kant's Transcendental Idealism." In *SEP*, Mar. 4, 2016. https://plato.stanford.edu/archives/spr2021/entries/kant-transcendental-idealism/.

Steiner, George. *Martin Heidegger*. Chicago: University of Chicago Press, 1989.

Steinkraus, Warren E., and Michael H. Mitias. *Taking Religious Claims Seriously: A Philosophy of Religion*. Amsterdam: Rodopi, 1998.

Steinmetz, Sol. *Semantic Antics: How and Why Words Change Meaning*. New York: Random House Reference, 2008.

Strehl, Manuel. "U+2621 Caution Sign." Codepoints, n.d. https://codepoints.net/U+2621.

Sundas, Binu. "Unit-2 Theory and Research." eGyanKosh, Mar. 2021. https://egyankosh.ac.in//handle/123456789/73800.

Suzuki, Daisetz T. *The Essentials of Zen Buddhism: Selected from the Writings of Daisetz T. Suzuki*. Santa Barbara, CA: Greenwood Press, 1973.

"Synopsis of Georg Cantor, 'Fondements d'une théorie générale des ensembles' ['Foundations of a General Theory of Sets']." Concept and Form, 1969. http://cahiers.kingston.ac.uk/synopses/syn10.3.html.

Tagore, Rabindranath. *The Complete Works: Poetry, Plays, Novels, Short Stories, Essays and Lectures, with Autobiography and Letter*. Prague: e-artnow, 2020.

———. *The English Writings of Rabindranath Tagore: A Miscellany*. Edited by Sisir K. Das. 8 vols. New Delhi: Sahitya Akademi, 1994.

Taylor, Ted. "Nevada Test Site Downwinders." Atomic Heritage Foundation, July 31, 2018. https://www.atomicheritage.org/history/nevada-test-site-downwinders.

Tegmark, Max. "The Wisdom Race Is Heating Up." In *Know This: Today's Most Interesting and Important Scientific Ideas, Discoveries, and Developments*, edited by John Brockman, 204–7. New York: HarperCollins, 2107.

Teilhard de Chardin, Pierre. *The Phenomenon of Man*. New York: HarperCollins, 2008.

Tester, Keith. "Between Sociology and Theology: The Spirit of Capitalism Debate." *Sociological Review* 48 (2000) 43–57.

Thomas, Christelle. "Why Do Abstract Art Images Make Us Feel So Good?" IdeelArt, July 27, 2017. https://www.ideelart.com/magazine/abstract-art-images.

Thomas, Norman L. *Modern Logic: An Introduction*. New York: Barnes and Noble, 1966.

Thompson, Mel. *Philosophy of Religion*. London: Hodder Arnold, 2003.

———. *Teach Yourself Philosophy*. London: John Murray, 1995.

———. *Teach Yourself Philosophy for Life*. London: Hachette, 2017.

———. *Through Mud and Barbed Wire*. Scotts Valley, CA: CreateSpace, 2017.

———. *Understand Philosophy*. London: Teach Yourself, 2010.

Thornhill, Chris. "Karl Jaspers." In *SEP*, June 5, 2006; revised July 17, 2018. https://plato.stanford.edu/entries/jaspers/.

Tillich, Paul. *The Courage to Be*. New York: Collins, 1963.

———. *Reason and Revelation, Being and God*. Vol. 1 of *Systematic Theology*. Chicago: University of Chicago Press, 1973.

Toulmin, Stephen. *Cosmopolis: The Hidden Agenda of Modernity*. Chicago: University of Chicago Press, 1992.

Travis, Stephen H. "Hope." In *New Dictionary of Theology*, edited by Sinclair B. Ferguson and David F. Wright, 321–22. Downers Grove, IL: InterVarsity, 1988.

"The Tribal Wisdom of the Dakota Indians." *Guardian*, Nov. 26, 1999. https://www.theguardian.com/money/1999/nov/26/workandcareers1.

Uckelman, Sara L. "William of Sherwood." In *SEP*, May 11, 2016; revised Aug. 25, 2020. https://plato.stanford.edu/entries/william-sherwood/.

Vaihinger, Hans. *The Philosophy of "As If": A System of the Theoretical, Practical and Religious Fictions of Mankind*. Translated by Charles K. Ogden. London: Routledge and Kegan Paul, 1925.

Van Leeuwen, Richard. *Narratives of Kingship in Eurasian Empires, 1300–1800*. Leiden, Neth.: Brill, 2017.

Varzi, Achille. "Boundary." In *SEP*, Feb. 9, 2004; revised Oct. 10, 2013. https://plato.stanford.edu/entries/boundary/.

Vesey, Godfrey, and Paul Foulkes. *The Collins Dictionary of Philosophy*. London: Collins, 1990.

Von Hayek, Friedrich A. "The Pretence of Knowledge." Nobel Prize, Dec. 11, 1974. https://www.nobelprize.org/prizes/economic-sciences/1974/hayek/lecture/.

Wagner, Roy. *Symbols That Stand for Themselves*. Chicago: University of Chicago Press, 1986.

Walschots, Michael H. "Moral Sense Theory and the Development of Kant' s Ethics." PhD diss., University of Western Ontario, 2015.

Ward, Graham. *A Postmodern God: A Theological Reader*. Hoboken, NJ: Blackwell, 1997.

Warner, Charles D. "Voltaire to Rousseau." In *A Library of the World's Best Literature—Ancient and Modern*, edited by Charles D. Warner, 38:15,484–90. New York: Cosimo, 2008.

Watson, John B. *The Ways of Behaviorism*. New York: Harper, 1928.

Weatherford, Jack. *The History of Money*. New York: Crown, 2009.

Webb, Richard, ed. *How Numbers Work*. London: John Murray Learning, 2018.
Weber, Max. *From Max Weber: Essays in Sociology*. Hove, UK: Psychology, 1991.
———. *The Protestant Ethic and the Spirit of Capitalism*. Translated by Stephen Kalberg. New York, Routledge, 2001.
Weil, Simone A. *Gravity and Grace*. Translated by Emma Crawford. New York: Putnam, 1952.
———. *Late Philosophical Writings*. Translated by Eric O. Springsted and Lawrence E. Schmidt. Notre Dame, IN: University of Notre Dame Press, 2015.
Weinrich, Peter. "Identity Structure Analysis." In *Analysing Identity: Cross-Cultural, Societal and Clinical Contexts*, edited by Peter Weinreich and Wendy Saunderson, 7–76. Abingdon, UK: Routledge, 2013.
Wellman, Carl. "Rights." In *The Cambridge Dictionary of Philosophy*, edited by Robert Audi, 695–96. Cambridge: Cambridge University Press, 1996.
Wells, Samuel. *What Anglicans Believe in the Twenty-First Century*. London: Cassell, 2000.
Whitehead, Alfred N. *Process and Reality*. New York: Simon and Schuster, 2010.
Whitmore, William H., et al. *Mother Goose's Melody*. London: Newbery, 1760.
Whorf, Benjamin Lee. *Language, Thought, and Reality: Selected Writings of Benjamin Lee Whorf*. Edited by John B. Carroll et al. Cambridge, MA: MIT Press, 1956.
Wilford, Paul T. "A Review of Ferrarin, Alfredo, *The Powers of Pure Reason: Kant and the Idea of Cosmic Philosophy*." *Kantian Review* 20 (2016) 325.
Will, Clifford M. "Einstein's Relativity and Everyday Life." Physics Central, n.d. https://physicscentral.com/explore/writers/will.cfm.
Williams, Raymond. *Resources of Hope: Culture, Democracy, Socialism*. London: Verso, 1989.
Wilson, Edward O. *Half-Earth: Our Planet's Fight for Life*. New York: Norton, 2016.
Wilson, Kelly G. "Some Notes on Theoretical Constructs: Types and Validation From a Contextual Behavioral Perspective." *International Journal of Psychology and Psychological Therapy* 1 (2001) 205–15.
Wittgenstein, Ludwig. *In Search of Meaning: Ludwig Wittgenstein on Ethics, Mysticism and Religion*. Edited by Ulrich Arnswald. Karlsruhe, Germ.: KIT Scientific, 2009.
———. *On Certainty*. English and German ed. New York: Harper and Row, 1972.
———. *Philosophical Investigations*. Edited by Gertrude E. M. Anscombe et al. Translated by Gertrude E. M. Anscombe. Hoboken, NJ: Wiley, 2010.
———. *Tractatus Logico-Philosophicus*. Translated by David F. Pears and Brian McGuinness. London: Psychology, 2001.
Wolchover, Natalie. "A Different Kind of Theory of Everything." *New Yorker*, Feb. 19, 2019. https://www.newyorker.com/science/elements/a-different-kind-of-theory-of-everything.
———. "How Gödel's Proof Works." Quanta Magazine, July 14, 2020. https://www.quantamagazine.org/how-godels-incompleteness-theorems-work-20200714/.
Wood, W. Jay. *Epistemology: Becoming Intellectually Virtuous*. Downers Grove, IL: InterVarsity, 1988.
Zangwill, Nick. "Aesthetic Judgment." In *SEP*, Feb. 28, 2003; revised Jan. 28, 2019. https://plato.stanford.edu/entries/aesthetic-judgment/.
Zweynert, Astrid. "Does Modern Slavery Need a New Definition?" World Economic Forum, Dec. 1, 2015. https://www.weforum.org/agenda/2015/12/does-modern-slavery-need-a-new-definition/.

Subject Index

abandonment, 270-71
  defined, 305
  as deconstruction, 165, 272
  interior decorator (illustration), 271
  as reordering, 271-272
  total, 165, 269
  See also awakening; truth values
abstract art
  See art, abstract
abstraction, 93-95, 200-202
  a continuum, 202
  always falsifies, 191-92
  mathematical, 185
  scientific, 200, 301
  See also individuation
actions, 151-52
  the greengrocer (example), 79, 82
  joining the resistance (example), 79
  non-violent direct action, 154
  predetermined, 165
  problem of action, 281-82
  spontaneous, 74
  for unclear reasons, 82
  See also ethics, personal; motivation; others
algorithms, 215-16
  limited in scope, 216
  vs. reality, 186-87
  used to fill gaps, 255
  See also mathematics; theory

anaphoras, 39-40
  defined, 305
  See also bridging inferences; words, subatomic
appearances, 18
  easily distinguished, 18
  and things in themselves, 14
  See also noumena; phenomena
arrangement of things, 71-75
  cat flap (illustration), 83
  esthetic, 90, 92, 93
  ethical, 84
  functional, 83
  and *ought*, 178-79, 197, 233, 242
  See also others; relations
art, abstract, 93-95
  vs. figurative, 95
  and human works, 94
  stepping onto a bus (example), 95
  transcendence, 95
  and urbanization, 94
  See also art, figurative; esthetics
art, figurative, 95
  vs. abstract, 95
  and preservative instinct, 96
  sets the stage for action, 93
  See also art, abstract; esthetics
atomic words, 28-29
  defined, 305
  See words, atomic

Subject Index

attention, 251–52
  defined, 250
  See also consciousness; reason
awakening, 161–68
  defined, 305
  touching a painting (illustration), 163
  and religious conversion, 165
  and wokeism, 162
  See also truth claims
awareness
  See attention

balance, 58–59, 90
  defined, 305
  See also education; esthetics; ethics, personal
behavior, 71–72
  aging mobster (illustration), 152
  happy family (illustration), 151–52
  an issue of orientation, 162
  rooted in relations, 61, 71
  See also awakening; behaviors, unvirtuous; behaviors, virtuous
behaviors, virtuous, 150–51
  the mean, 157
  teaching of, 258
  See also ethics, theories of
behaviors, unvirtuous, 150–56
  defined, 310
  forms of, 153–56
  house on sand (illustration), 152
  parking lot (illustration), 153–55
  See also psychology
belief
  in causality, 236–38, 248
  in God, 277, 294, 309
  and history, 241
  See also faith; truth claims
big data
  See data
binary
  See Boolean
body, the
  ease of body, 292
  hardships, 281
  is instrumental, 72
  See also mind-body dualism

bondage, 21–27
  defined, 305
  Hey diddle diddle (rhyme), 23
  hotel lobby (illustration), 27
  See metaphysics, unacknowledged; words, subatomic
Boolean, 177–78
  computer bits, 4
  and *is* and *ought*, 177–78
bridging inferences, 38–40
  defined, 305
  See also anaphoras; words, subatomic
built environment, 147–49
  defined, 305
  See also environment, built

calculated
  ethics, 86
  life, 219, 240
  See also free will; mathematics
calling, 124–27
  defined, 305
  being and calling, 126, 305
  benefits all, 126–27
  employment as, 125–28
  impediments removed, 125
  See also employment
capitalism, 216
  economic efficiency, 216
  ruthless activity, 216
  See also systems, failed
causality, 232–40
  all-inclusive, 13, 224, 234
  causes of war (example), 233, 239
  is created, 235, 238, 239
  and God, 275–76, 282–84
  marbles on a scale (illustration), 234
  toppling dominoes (illustration), 232
  See also determinism; free will
centers, 25–27
  to be avoided, 66–67
  exclude, 67
  and origins, 25, 36, 163
  and prayer, 26
  in theology (examples), 67
  vs. things and relations, 68, 299
  See also single thought

Subject Index

certainty
   and history, 239, 245
   and religious history, 242–43
   and scientific knowledge, 182
China, Chinese
   See Eastern philosophy
choice
   assumption of, 253
   of cause and effect, 238, 277
   freedom of, 248, 250
   subconscious, 82
   of variables, 242–43
   See also causality; determinism; free will
Christianity
   Christian Platonism, 24
   and sovereignty, 268–69
   See also Name Index; religion; secularism
citizens
   citizen's complaint (example), 116–17
   contract with the state, 271–73
   equal before law, 130
   identity politics, 132
   See also democracy; political systems; syntocracy
codes
   of ethics, 72, 85–88
   law code, 108
   semiotic, 75, 98, 113, 130, 218, 258
coercion, institutional, 129–34
   defined, 305
   excessive reliance on, 132
   imposition of relations, 129–30
   instruments of state, 129–34
   See also coercion
coherence, of objects, 20, 38, 226
   is illusory, 277, 303
   muddied thinking, 237
   See also things, dissipated
community, 113–14
   common compassion, 113–15
   Homo reciprocans, 87
   personable public servants, 114
   values over individualism, 112
   See also groups; political systems

complexity
   of God's activity, 283–84
   of the human sciences, 221
   vs. minimum state (illustration), 109–10
   of nature, 143–46
   of others 74–75, 218
computers
   big data, 215
   and incompleteness theorem, 194
   limitations, 62, 145
   oversimplify reality, 203
   See also Boolean
concepts, 18–20
   defined, 305
   big concepts, 239–40
   are things, 19, 71
   See also high-level; low-level; things
connotation, 41–42
   defined, 306
   See also defintions; denotation; synotation
consciousness, 248–51
   and attention, 248–49
   cut on arm (illustration), 250
   after death, 287–88
   hierarchy of the senses, 251
   phylogenetic scale, 247
constructs, 15–18
   constructed meaning, 294
   falling ants (example), 47
   things not perceived, 16–17, 31, 192
   units of math, 190
Continental philosophy
   Bourbaki group, 231
   Enlightenment, 302
   Frankfurt school, 96, 129, 220
   French *philosophes*, 111
   Hegelian-style, 198
   Post-Jungians, 94
   See also Name Index
continuation, 287–89
   defined, 306
   afterlife, 285–90
   hunting grounds (example), 287
   See also death

contradiction, 254–56
  defined, 254, 309
  and loss of hope, 9
  in philosophy, 5–8
  reason and, 254–56, 309
  See also contradiction, principle of
contradiction, principle of
  and motivation, 71
  and unvirtuous behavior, 152
  x and not-x, 203, 242
control, 62–65
  of death, 286
  limitations of, 62–65
  over nature, 143–44
  over others, 158
  strategic rationalism, 55
  See also scientific method; systems, failed
creation, 238–39
  of causality, 238, 249
  of centers, 67, 69, 99, 164
  God's, 276
crime and punishment, 132–34
  capital punishment, 134
  cost and benefit, 133
  crime is far-reaching, 112, 216
  provides context, 134
  purpose of punishment, 129, 133
  See also behavior, unvirtuous; coercion, institutional
crisis, 9–10
  emotional and spiritual 164
  epistemic, 9, 143, 146
  global, 54, 143, 209
  of real vs. unreal, 15
critical theory, 157–60
  defined 306
  and emancipation, 157
  and environment, 160
  and information, 158–59, 306
  origins in antiquity, 157–58
culture
  inhabits grammar, 100
  is transmitted through language, 76, 77
  is ordinary, 135
  pluralism, 137–38
  See also inheritance

dangers
  artificial intelligence, 216
  excessive control, 65
  human limitations, 62–65
  mathematics, 191, 202
  reason, 253
  religion, 267, 273
  science, 206, 211–14
  worldviews, 10, 163
data, 214–16
  big data, 69, 214–16
  reduces nature, 214
  subjective, 252
death, 285–90
  defined, 306
  alters behavior, 285, 286
  fear of, 286
  severance of relations, 285–88
  See also continuation
deconstruction
  See abandonment
definitions, 41–42
  a car (example), 229
  basic to causality, 235
  describe priority relations, 42
  reach into all of reality 226, 248
  See also connotation; denotation; synotation
democracy, 107–109
  defined, 306
  friends move a table (illustration), 129–30
  seeks interrelatedness, 107
  shortcomings, 108–109
  See also political systems; syntocracy
denotation, 41–42
  defined, 306
  See also defintions; connotation; synotation
de-selection of facts
  envionmental, 59, 242
  scientific, 188, 212, 242
  social and historical, 242
  See also critical theory
determinism, 232–40
  out of the question, 236
  rests on As and Bs, 236–37
  See also free will

Subject Index

disjoint
   all models are wrong, 192, 203, 270
   between expectation and reality 279
   between mind and reality 93
   and sense of unreality, 249
   triggers visceral reaction, 74, 151–52
   See also consciousness; mind–body dualism
dissipated words, 226–31
   defined, 306
   See things, dissipated; words, dissipated

Eastern philosophy, 6, 12
   Confucianism, 114
   Eastern mysticism, 298
   Faija school, 85
   Hundred Schools of Thought, 6
   See also Name Index
economics, 122–28
   defined, 306
   and commerce, 214–16
   the commute (illustration), 126–27
   constitutional guarantees, 128
   major approaches, 122–23
   market forces, 220
   new self-adjusting, 126–27
   See also employment
education, 258–63
   defined, 306
   deficient knowledge, 258
   holistic, 221, 258–63
   individualized attention, 144
   political 132, 262
   silent forces, 257
efficiency, 218–20
   economic, 216, 218–20
   legitimation for human action, 217–19
   musical grandfather (illustration), 220
   science of scarcity, 214
   See also capitalism
emergent features, 11, 54–55
   field of wheat (illustrations), 55–56, 70
   meta-features, 53, 61, 162
   and metaphysics, 300
   See also ethics, personal

emotion, 71–74, 113–14
   affective information, 159
   emotional dysfunction, 155
   relations and, 72
   See also disjoint; passion
employment, 122–28
   and economic theory, 122
   minimum wage, 127–28
   oppressive, 123–25
   and social liberation, 126–27
   two acrobats (illustration), 128
   wealth gap, 110
   See also calling
endless expanse, 53–61
   defined, 306
   new view of reality, 8, 12, 33, 56
   precludes centers, 97
   reveals meta-features, 53, 55
   source of ethics, 53, 57
   See also emergent features
English
   100 most common nouns, 15
   dominoes (illustration), 33
   grammar, 99–101
   subject and predicate, 34–37
   transmits culture, 76–77
   See also inheritance
environment, built, 147–48
   defined, 305
   and art, 94
   invasiveness of, 147
   mountain pass (illustration), 147
   separation from natural, 149
   See also environment, natural
environment, natural, 143–49
   defined, 308
   big data excludes, 215
   environmental ethics, 143–49
   epistemic crisis, 143, 146
   exclusion of, 59, 67, 242
   management of, 144, 148
   a single pond (illustration), 145
   See also critical theory; environment, built
epistemology, 14–18
   all things are real, 18–19
   all things declared fictions, 15
   epistemological imperialism, 221
   reading a novel (illustration), 246

equality
  before the law, 130
  common humanity, 142
  of information, 116–21, 302
equality benefits, 140–41, 306
  defined, 306
  and absolute rights, 113
  and freedom of religion, 140
  See also critical theory; identity
essence, 33–37
  concealed, 77
  flees away, 225, 230
  of life 240
  See also synessence
esthetics, 89–96
  defined, 90, 306
  couch in a room (illustration), 90
  and ethics, 89–93
  and interpretation, 60
  regulation of art, 92
  See also art, abstract; art, figurative
ethics, personal, 51–103
  defined, 306
  and digital interactions, 216
  grounded in physical reality, 256
  moorings lost, 53–54
  occasional opacity, 79–83
  ten ethical maxims, 56–70
  See also emergent features; maxims, ethical; pseudo ethics
ethics, public, 105–65
  avoids extremes, 157
  and employment, 122–28
  and environment, 144–46
  and freedom, 111–12
  incentives, 130
  political, 107–14
  See also employment; information; political systems
ethics, theories of, 54–55
  interrelation, 84–88
  rational vs. natural, 85–86
events
  defined, 306
  de-selection of, 242
  historical, 241
  providential, 245, 274, 277
  are things, 47

everything
  complete in everything else, xiii
  connects to everything else, xvi, 11–12, 107, 274
  influences everything else, 235, 240
  mulberry tree (illustration), 13
  return to a premodern outlook, 277
  rising awareness of, 11–13, 276
  See also holism
expectations, 71–76
  gold foil experiment (example), 183
  and motivation, 71–74
  and *ought*, 171–172
  and political legitimacy, 131–32
  and scientific discovery 183
  See also motivation; behavior, virtuous; behavior, unvirtuous
experience, 41–42
  of causality, 236
  of God, 282
  hike by moonlight (illustration), 41
  language games, 41, 272

facts, 175–78
  defined, 306
  in education, 260
  historical, 242–43
  are illusory, 230–31, 301
  position of table (illustration), 176
  are selected, 187, 233
  the Scots keep unicorns (example), 176
fact–value distinction, 53–54, 174–76, 218–220
  mirrors mathematical thinking, 196
  my house has moved (illustration), 178
  we can, and do, 174
  See also things–relations distinction
faith
  correlated with all truths, 245
  in God, 276, 277
  a leap in the dark, 244
  saving faith, 287
falsifiability, 206–7
  defined, 306
  in closed systems, 206
  See also invalidation

Subject Index

fear
  of death, 285–86
  and fundamentalism, 67
  See also hope
features
  See meta-features; emergent features
feeling
  See emotion; passion
first philosophy, 21–22
  defined, 306
  abandonment of, 165
  all possess, 21
  and behavior, 155
  See also metaphysics
focus
  and fringe 199
  low and high, 95
  people-focused politics, 108
  and reason, 225
force
  unlawful, 131
  unvirtuous, 150, 154–55, 158
  attractional force of words, 26
forms
  abstract, 5, 186
  of life, 41
  marked, 35
  Plato's forms, 5
  See also theory
foundations, 1–49
  defined, 307
  as context, 1
  things and relations, 164–65
  See also certainty
free will, 232–40
  defined, 307
  and mind, 247–48
  rice farmers (illustration), 237
  shot at a bell (illustration), 238
  unquestionable, 236–38
  See also causality; determinism
freedom, 111–13
  and accident, 240
  counterproductive, 111
  of information, 118, 160
  and productivity of thought, 22
  of religion, 140, 272
  subjective, 237
  See also bondage; free will; freedom

fundamentalism
  See fear
future, the
  and consequentialism, 84
  cut short, 287
  and hope, 283
  safety of religion, 273
  utilitarianism, 55

givens, 252–56
  basis of ethics, 256
  ceteris paribus, 186
  purpose of reason, 250, 252–56
goal
  of big data, 215
  of critical theory, 157
  of democracy, 109
  of ethics, 88, 113
  of psychology, 156
  of reason, 163
  of society, 110, 125
God, 275–84
  defined, 307
  and causality, 275–76
  must think it odd, 17
  possible vs. necessary, 278, 284
  reality or fiction, 17–18
  See also God's action; religion
God's action, 280–84
  always good, 284
  beyond causality, 279–80
  commissioning editor (illustration), 278
  complexity of, 284
  first cause, 276
  God above God, 295
  providence, 277, 279–84
good
  the beautiful and the, 89
  common good, 112, 133, 141
  good deeds, 165, 283
  possibility of the, 284
grammar, 99–101
  building materials (illustration) 20
  conjugations and declensions, 99–101
  no essence, 100–101
  reflection of a culture, 100
  in space and time, 100
  subject and predicate, 34–37

Greek philosophy, ancient, 12, 34, 160
    Aristotelianism, 298
    Cretandom, vi
    Epicureanism, 298
    Ionian and Italian schools, 6
    Platonism, 298
    Stoicism, 293, 298
    See also Name Index
groups
    consensus, 130
    and identity, 139
    individual prior to, 112
    like-minded, 111
    minority groups, 132

happiness
    meaning of life, 292
    understanding one's limits, 64
harm, 169–70
    defined, 307
    collective harm, 269–70
    inherited thinking, 269
    and ownership, 270
    See also awakening; sin; truth claims
hegemony
    of causality, 240
    dismantling, 157, 160
    Enlightenment hegemony, 302
    of money, 124
    of physics, 221
health
    body and mind, 259
    as liberation, 126–27
    of society, 120
Hinduism
    and sovereignty, 268–69
    See also Name Index; religion
high-level, 3–5
    defined, 307
    windmill (illustration), 299
    See also concepts; low-level
history, 241–46
    defined, 307
    and causality, 241
    criteria for judging, 243–44
    de-selection of events, 242
    no total knowledge, 64, 244
    personal history (illustration), 241
    See also religion

holism, 10–13
    defined, 307
    in economics, 126–27
    in education, 259
    and environment, 144
    in ethics, 56–57
    in society, 112, 142
hope, 280–84
    false hope, 289
    in God, 283–84
    personal vs. social, 283–84
    spontaneous in youth, 281
    unconditional, 284
hubris, 210–16
    defined, 307
    doomed, 210
    and mathematics, 202
    of progress, 64–65
    of science, 210–216
human nature, 77
    abuses of government, 115
    wretched, 304
    See also inheritance

ideas, 3–5
    defined, 307
    central ideas, 68
    dichotomous, 34
    of first importance, 3
    infiltrate our minds, 3–4
    See also concepts
identity, 135–42
    defined, 135, 307
    conflicts of, 137–39
    driven by necessity, 137
    immutable, 136, 138, 140
    may be harmful, 141
    is ordinary, 135
    roles, 136–37, 141
ideology, 262–63
    and education, 263
    political education, 132
    and relations, 262–63
    See also truth
individualism
    and community, 112, 128
    counterproductive, 111–12, 127
    and death, 289
    and meaning, 291
    See also community; groups

individuation, 197–201
    defined, 307
    focus and fringe, 199
    from an undifferentiated stream, 198
    units and complexes, 190–91
    yelping dog (example), 199
inequality 140–41
    and critical theory, 157–58
    and economics, 128, 110
    and identity, 139
    and social media, 111
infinite, 10–13, 62–65
    defined, 307
    infinite regress, 49, 225
    and natural environment, 94
    things and relations, 47, 62, 68, 270
    See also limitations; relations
information, 115–21
    defined, 307
    access to, 118
    citizen's complaint (illustration), 116–17
    dangers of withholding, 115–16, 158
    Fourth Estate, 120
    government presses, 118
    mediated, 119
inheritance
    of culture, 77, 269
    of harmful behavior, 269, 304
    of thought, 270
    through words, 40, 310
    See also culture; sin
instinct, 25–27
    actions, 74
    ethics, 80, 218
    philosophical thinking, 26
    instincts in society, 124
instruments
    of civil justice, 133
    others become, 219
    of self-destruction, 212
    See also body, the; instruments of state
instruments of state, 129–34
    defined, 307
    See also coercive institutions; instruments

interrelated world
    See awareness; everything
intuition, 207–9
    linguistic, 39
    moral, 54, 61, 217
    scientific, 207
    See also scientific method
invalidation, 203–8
    defined, 307
    vs. falsification, 206–7
    in a global system, 206–7
    Ohm's law (example), 185, 206
    See also falsifiability
is-ought distinction, 53–54
    defined, 207
    is cannot become *ought*, 174
    is imples *ought*, 176–77
    See also fact–value distinction
Islam
    Greater Sin, 161
    Islamic art, 93
    and sovereignty, 268–69
    the way appears, 88
    See also Name Index; religion; secularism
isolation, 203–9
    of causes, 235
    of mathematics, 191
    of mind, 248–49
    of regions of relations, 95
    of science, 203–9
    See also individuation

judgment, 76–78
    based on information, 158
    and critical theory, 157
    esthetic, 89–90
    moral, 54–55
    value judgments, 178, 300
    See also history; value
justice
    criminal justice system, 134
    rehabilitative, 133
    retributive 133–34
    a service of the state, 127
    worlds beyond our own, 110–11

Subject Index

knowledge, 214
  deficient in itself, 214, 258
  exclusionary, 145, 187, 212, 258
  and imagination, 207–208
  and social competence, 221
  unity of, 245
  See also science

labels, 15–20
  no *primarius* or *secundus*, 66
  and synotation, 30
  things and, 18–20
  are words, 20
  See also concepts
language
  defined, 307
  distorts and misrepresents, 304
  dominoes (illustration), 33, 163
  embodies worldview, 76–77
  and mind, 248–50
  new theory of language, 36
  and relations, 42, 226
  See also grammar; structural linguistics; words
languages, 197–202
  buying rope (illustration), 197
  language is languages, 11
  three kinds of, 197
laws, 107–9
  constitutional, 117
  of noncontradiction, 71, 203
  statutory, 108
  wisely chosen, 130
  See also theory
leadership, 113–14
  obduracy, 108
  requirement of, 114
  transformational, 108
liberal
  arts, 259
  post-liberal theology, 265
  priorities, 111
liberation, 28–37
  as abandonment, 78, 271
  and critical theory, 157
  language unbound, 24, 32–33
  social, 126

limitations, 21–27, 62–70
  human, 22 62–70, 159, 270, 299
  knowing one's, 60, 62–63, 145, 262
  of language, 23, 37, 42, 70, 300
  limit situations, 163
  regrets and impotences, 304
  of theocracy, 274
  See also language; theory
linguistic features
  See anaphoras; bridging inferences; grammar
linguistic system
  defined, 308
  vs. human activity, 32
  See also language; structural linguistics
logic
  is dumb, 175
  and identity, 137
  and reason, 83, 254
  RFL–class, xi
love, 256–57
  and causality, 239
  holistic quality, 257
  humanism 85
  love restored, 300
  See also community
low-level, 3–5
  defined, 308
  machine code (illustration), 4
  professor and student (illustration), xv
  See also concepts; high-level

machines
  inscrutable, 119
  machine learning, 215
  means of control, 62
  and mind, 249
mathematics 189–96
  defined, 308
  an abstraction, 201–2
  carved up, 192–93
  idealized, 187, 191–193
  mushroom sauce (illustration), 191
  propagation of errors, 194
  units subjective, 190–91
  and W, the whole, 195
  See also fact–value distinction; individuation

Subject Index 341

maxims, ethical, 56–70
  defined, 308
  and freedom of information, 116
  and nature, 144–45
  and physical science, 212, 216
  ten maxims listed, 69–70
  and unvirtuous behavior, 153–54
  See also emergent features; ethics, personal
meaning, 291–96
  defined, 308
  beyond bounds, 294
  loss of, 291, 294
  motivates, 294–95
  the rose garden (illustration), 293–94
  is social, 291
  sociology explains it (illustration), 295–96
  See also words
meta-features, 38–44
  defined, 308
  See also emergent features
metaphysics
  defined, 308
  awful consequences, 10
  everything, briefly, xvi, 10–11, 33, 296
  incredulity towards, 9
  and religion, 268, 300
  surveyed, 297–302
  See also metaphysics, unacknowledged; philosophy
metaphysics, unacknowledged, 25–27
  defined, 310
  periodic table (illustration), 23
  powerfully controls, 25–27
  roll of the dice (illustration), 29
  See also language
metaphysicians
  ambitions, 21, 26
  of modernity, 302
  self-importance, 7–8
  system-builders, 255
  See also philosophers
method
  of this book, xv
  See also scientific method

mind, 247–51
  and behavior, 61
  difficult to shift, 138
  mind-independent, 5
  and reality, 17, 62–63, 249
  See also limitations; mind–body dualism
mind–body dualism, 247–51
  defined, 308
  glass of milk (illustration), 247
  isolation of the mind, 248
  and materialism, 247
  related to individuation, 249
  See also mind
moral
  defined, 308
  See also ethics
motivation
  defined, 308
  and conceptions of the world, 151–52
  dog and bowl (example), 72
  and God, 284
  grand master (illustration), 74
  loss of, 281–83, 285
  moving the table (illustration), 73, 74
  and principle of contradiction, 71
  rational, 282
  See also others

names, 15–20
  a flat ontology, 19
  individuation, 198
  recall things, 19, 28
  See also labels
nation
  See political systems
natural environment
  See environment, natural
necessity
  of afterlife, 290
  of God, 278
  mother of reason, 254
  social, 137
  See also possibility
negative ethics, 62–70
  defined, 308
  See also positive ethics

noumena, 14
    and phenomena, 14
    See also things in themselves
number
    critical to metaphysics, 4–5, 46
    and human limitations, 62
    of things and relations, 46–48
    and totality, 45
    we are just a number, 219

objects
    See things
ontology, 14–20
    defined, 308
    See also epistemology; things; words
opposites
    See oppositions
oppositions, 80–82
    clarify, 80–81
    types of, 80–81
    of virtue, 151
oppression, 157–60
    of the environment, 160
    identification of, 159
    and loss of information, 158–59
    and uptake of information, 116–18
optimism
    irrational, 210
    of science, 214
    social optimists, 64
others, 73–75
    cannot be reduced, 75, 257
    exclusion of, 67, 111, 293
    game of fast chess (illustration), 74
    and meaning, 291
    problem of, 73–75
    rapport, 75, 87, 113–14, 260
    sitting on the table (illustration), 74
    See also community
ought, 169–73
    defined 169–70
    Egyptian *nfr* (example), 194
    no *ought* in nature, 146
    not a moral word, 173
    variations of meaning, 170–71
    See also expectations; is–ought distinction

ownership, 270
    defined, 308
    Confiteor Deo, 304
    See also property; truth claims

paradigm shift, 208–9
    Global Positioning System (GPS), 182
    Newton to Einstein, 181
    and the scientific method, 208
passion, 71–73
    and reason, 72–73
    reason in, passion out, 73
    See also emotion
people, persons
    See others; codes
perception
    of the senses, 16
    unity and diversity, 235
    See also appearances
phenomena, 14
    and noumena, 14
    See also appearances
philosophers, 3–8
    bound by rules, 26
    bounded view, 12
    confusion of, 7–8, 21
    our debt to, 297
    See also Name Index; metaphysicians
philosophy, postmodern
    centers rejected, 67
    characteristics, 302
    in crisis, 5, 9
    See also philosophy of relations
philosophy of relations
    defined, 308
    exits a fixed frame, 68
    Grand Unified Philosophy, 1, 298
    vs. modern philosophy, 301
    a philosophy of relativity, 299
physics, physical science
    See also science; scientific method
pluralism, social, 137–39
    and absolute rights, 136–40
    affirms value of all, 142
    challenge to order, 138
    See also critical theory

Subject Index 343

political systems, 107–114
  balanced relations, 109, 115–16
  democracy, 107–9
  open society, 115, 120–21, 273
  personal liberty, 111–12
  philosopher kings, 109
  political legitimacy, 131
  syntocracy, 109
  See also freedom; information; secularism
positive ethics, 53–61
  defined, 308
  See also negative ethics
positive religion
  defined, 308
  and law code, 108
  rejected, 271
possibility
  of alternative reality, 277–80
  of God, 278
  See also God; necessity
powers
  See limitations
predicate
  See subject
prediction
  of consequences, 206
  failure of, 212–13
  of human behavior, 75
  scientific, 180
  See also mathematics
preservative instinct
  in art, 96
  in language, 96
  and meaning, 296
pride
  See hubris
productivity, 21–22
  of language, 22
  of thought, 22
  See also freedom
property, 112–13
  essential to progress, 113
  and human flourishing, 113
  not the *summum bonum*, 112
  rights, 112–13
providence
  See God's action

pseudo ethics, 217–21
  defined, 309
  capsized ship (illustration), 221
  dehumanizing, 219
  fills a vacuum, 217
psychology, 155–56
  and disturbance of identity, 136
  goal of, 156
  and mental dysfunction, 155–56
  See also goal
punishment
  See coercion, institutional; crime and punishment
purpose
  of employment, 125
  of life, 292
  is preexistent, 252
  See also goal; meaning

reality, 14–20
  defined, 309
  no direct access, 14
  rules of chess (illustration), 279
  Thomas theorem, 238
  See also epistemology; possibility; things
reason, 252–57
  defined, 309
  and attention, 253
  bridges gaps, 61, 254–56, 301
  buzz of a mall (illustration), 253
  house without a roof (illustration), 254
  last judge and guide, 253
  responds to contradiction, 253–54
  See also givens; love; passion
reconciliation, 297–302
  Eastern and Western, 298
  ethics and esthetics, 154
  fact and value, 176
  free will and determinism, 236
  individual and group, 127
  ordinary and scientific language, 202
  reason and passion, 73
  religion and reason, 269

reduction, 233–40
    of events of history, 239
    of information, 120
    of nature, 146
    ontological, 237
    rejected, 56
    of relations, 274
regions of relations, 56–57
    defined, 309
    influence behavior, 56
    subsets of all relations, 49
regulation
    of the arts, 92
    regulators and ombudsmen, 117
    of scientific pursuits, 205
relatedness
    See everything
relations, 10–13, 21–37
    defined, 309
    boundless, 305
    broad vs. narrow, 56, 152, 161
    and ethics, 53–165
    kinds of relations, 47
    thunder (illustration), 48
    See also everything; things; words
relations, deeper, 225–31
    applicator (illustration), 228–29
    end in nothing, 227, 230
    fractal image (illustration), 225
    a seamless web, 226
    See also words, dissipated
religion, 267–74
    defined, 309
    freedom of, 140, 271
    history and, 244
    and secular law, 272–73
    supernatural assumptions 273–74
    test of truth, 245–46
    truth claims, 268–74
    See also God; positive religion;
        secularism
rights
    absolute, 136, 138–39
    benefits of, 113
    and compassion, 85
    individual, 111–12
    property, 113–14
    See also equality benefits

school
    See education
science, 179–88, 203–9, 216
    defined, 309
    benefits of, 203
    changes in beliefs about, 182
    demoted, 200, 220–21, 301
    fact and value, 179–88
    human sciences, 75, 181, 219–20
    privileging of, 218
    See also hubris; knowledge; theory
scientific method, 203–9
    defined, 309
    CCO metal detectors (illustration),
        207
    a closed system, 181
    described, 181, 204
    and ethical maxims, 212
    screens out, 204, 208
    on steroids, 215
    See also invalidation; paradigm shift;
        theory
secularism, 268–73
    origins, 268–69
    and truth claims, 269–73
    See also political systems; religion
self, 135–39
    and death, 289
    growth of inner self, 164
    know thyself, 60
    self-construal, 135
    self-interest, 56, 152
    See also identity
self-assembly, 25–27
    defined, 309
    and dichotomous ideas, 33–34
    of metaphysics, 26–27
semiotic codes
    See codes
side effects
    combustion engine (example), 63,
        191–92
    games in lighted rooms, 110
    insects in a bottle (illustration), 205
    multiply, 205, 213–14
    Test Able (example), 212
    unpleasant and unforeseen, 63, 191,
        213
    See also dangers

sin, 76–78
  concept expanded, 154
  original sin, 77, 270
  and redemption, 268
  sin or genes, 235
  See also culture; inheritance
single thought, 25–27
  and relations, 68
  and postmoden philosophy, 299, 302
  See also centers
social
  complexity, 109–10, 137–38
  hope, 283
  inequalities, 110, 158
  optimists, xix, 64
  See also employment; groups
South Asian philosophy
  heterodox and orthodox schools, 6
  Madhyamika school, 6
  Samkhya achool, 236
  See also Name Index
state
  See political systems; religion
structural linguistics, 25–27, 68, 175–76
  defined, 309
  and bondage of words, 25–26, 68
  liberation from, 33
  mosaic (illustration), 22–23
  See also linguistic system; words, subatomic
subject
  possibilities within, 36–37
  hides dangerous opposites, 37
  story about a woman (illustration), 35
  See also English
suffering
  awakening implies, 162
  and empathy, 87
  See also body, the
supernatural, 244–46
  defined, 309
  consolation, 281
  not excluded, 280
  and truth claims, 273–74
symbolism, 102–3
  defined, 309

expansive relations, 301
symbolic language, 102–3
synessence, 33–37
  defined, 309
  inclusive, not exclusive, 35–37
  vs. reduced reality, 235
  See also essence
synotation, 29–33
  defined, 30, 310
  euphemism (example), 31
  incorporates experience, 41
  and overlap, 31–33, 90
  the sheriff (illustration), 42
  See also connotation; definitions; denotation
syntocracy, 107–9
  defined, 109, 310
  and relations, 108
  See also democracy; political systems
systematization
  of information, 120
  involves omission, 242
  scoping the system, 188, 191
  See also systems, closed; systems, global
systems, closed, 185–88
  failed political systems, 163
  falsifiable, 206
  never truly closed, 185–86, 192–93
  railway system (illustration), 186–88
  See also concepts; systems, global
systems, global, 185–88
  different marks of truth, 244–46
  and externalities, 215
  and history, 242
  invalidation, 206
  science within, 301, 307
  See also invalidation; scientific method; systems, closed

teaching
  See education
technology
  defined, 310
  application of science, 205
  and engineering, 205
  See also knowledge; science

theory, 183–88
  competing explanations, 182
  not exhaustive, 193
  provisional and limited, 182
  scope of, 195, 203, 212
  See also critical theory; falsifiability; science
things, 45–49
  defined, 310
  bicycle track (illustration), 17
  are concepts, 19
  are contents of words, 20
  and events, 47, 235
  and relations, 3, 45–49, 183
  terms for, xix
  See also individuation; reality; relations
things, dissipated
  defined, 306
  and dissipated words, 229–30
  incomplete in themselves, 225, 228, 277
  See also words, dissipated
things in themselves, 15–18
  and relations, 175
  all things are real, 18
  things exist twice, 14
  See also appearances
things-relations distinction, 175–78
  defined, 305
  and past philosophers, 175–76
  See also fact-value distinction
time
  and cause, 232
  and space, 100
totality, 45–46
  defined, 310
  the environment, 149
  of one's self-construal, 135
  of relations, 45, 161, 271
  See also holism; maxims, ethical
totalizing tendencies, 64–65, 144–49
  and antichrist, 116
  to be avoided, 64, 139
  and environment, 144
  and perfectionism, 64–65
  unreachable mastery, 64

tracing relations
  defined, 49, 310
  and balance, 90
  incomplete, 63
  others trace relations, 74
  is to pursue truth, 57
tradition
  adaptable, 87
  and identity, 141
  and inductive process, 207
transparency, 116–21
  moderates power, 118–19
  scope of, 116–19
  sheds light, 118
  See also critical theory; information
transmission
  See inheritance
truth
  defined, 310
  abandonment of, 78, 161–62
  coherence and, 245
  incoherent friend (illustration), 245
  louder than words, 245
  and propaganda, 115
  truth of movies (illustration), 243
  See also tracing relations; words, subatomic
truth claims, 268–71
  defined, 310
  and awakening, 272
  credible in themselves, 245
  and secular state, 272

ultimate, the, 267–304
  final principles, 265
  ultimate meaning, 294
  ultimate motivation, 287
unacknowledged metaphysics
  See metaphysics, unacknowledged
unexpected, the, 71–73, 250–56
  and attention, 251
  and God, 283
  and motivation, 72, 151, 282
  pendulum (illustration), 250
  reason a reaction to, 253
  See also expectations; motivation; ought

Subject Index

universe
    of associations, 227
    greater than, 62
    intimately present, 211
university
    See education
unvirtuous behavior
    See behavior, unvirtuous

value, 53–54, 217–21, 174–78
    defined, 310
    fact–value distinction, 53, 176
    intrinsic value, 66
    value judgments, 178
    weighted words, 97–103
    See also fact; fact–value distinction; pseudo ethics
values
    See ethics; moral
view, 22–24, 54–56
    bird's-eye, 55, 68, 70
    bounded, 56
    narrow vs. broad, 56, 81, 152
    See also awakening
virtues
    See behaviors; moral

wages
    See employment
words, 38–44
    defined, 310
    contents, 20–23, 39, 226, 310
    not truth-neutral, 43–44
    tiddly-wink (example), 98–99
    tightening up meanings, 29
    weighted, 97–103
    See also grammar; labels; things

words, atomic, 28–29
    defined, 305
    invariant, 101
    minimum free forms, 28
words, subatomic, 38–44
    defined, 30, 309
    Aristotle's house (illustration), 39–40
    bivalence, 43
    inside, 30–31, 42
    proof of concept, 38–41
    stone (illustration), 30–31
    universe of associations, 227
    See also experience; metaphysics, unacknowledged; synotation
words, dissipated, 226–31
    defined, 306
    beget words, 248
    and common sense, 230–31
    H_2O (illustration)
    See also things, dissipated
work
    See employment
world
    has changed, 10–12
    in a grain of sand, 11
    in one's mind, 73, 82, 154–55, 270
    of relations 30–33, 82
worldviews, 21–22
    defined, 310
    and behavior, 156
    grouped in pairs, 9, 34
    hobbled, 163
    language embodies, 76
    See also dangers

Name Index

Adorno, Theodor, 94
Aho, James, 219
Ali, Rami A. el, 74
Anaxagoras of Clazomenae, 31
Andersen, Hanne, 182
Antoinette, Marie J. J., 343
Appelbaum, David, 212
Aquinas, Thomas, 272
Arachne, 210
Arcesilaus, 63
Archimedes, 208
Aristotle, xix, 5, 13, 21, 32, 39–40, 42, 43, 57, 150–51, 200, 234, 280, 292
Armstrong, David M., 247
Attenborough, David, 144, 145
Augustine of Hippo, 24
Aurelius, Marcus, 112
Austin, John L., 41
Averroës, Ibn R., 7
Ayer, Alfred J., 239,

Bacon, Francis, 24, 116, 226, 247, 248, 299
Bailey, Andrew M., 15
Bain, Alexander, 34
Barad, Karen M., 59, 221
Barnett, Lincoln, 197
Barrow, John D., 277
Barsalou, Lawrence, 35
Barthes, Roland, 12
Beard, Charles A., 242

Bell, Daniel, 205
Bense, Hermann, 29
Bentham, Jeremy, 133,
Bentley, I. Madison, 155
Bergland, Christopher, 281
Bergson, Henri–Louis, 205
Berkeley, George, 16, 280
Berlin, Isaiah, vi, 81
Berry, Thomas, 188, 210
Biase, Luca de, 213
Black, Max, 25, 40, 67, 95, 101, 230, 303
Blackburn, Simon, 15,26, 45, 138, 187, 233, 239, 250, 285
Blake, William, 11
Blöser, Claudia, 284
Bloom, Benjamin S., 259
Bloomfield, Leonard, 28, 305
Bohm, David J., 237, 261
Bohman, James, 157
Bonaparte, Napoleon, 276
Bonaventure (Giovanni di Fidanza), 272
Bonhoeffer, Dietrich, 146
Booth, Gordon T., 125
Bosanquet, Bernard, 111, 112, 256
Bowles, Samuel, 87
Box, George, 192
Bradley, Francis H., 6, 49, 239
Bridgman, Percy W., 187, 191, 194, 195, 228, 234
Britton, Karl W., 291
Bronowski, Jacob, 284

Name Index

Broome, John, 170, 173
Brouwer, Luitzen E. J., 190
Browne, Thomas, 87
Brunner, H. Emil, 6
Briullov, Karl P., 92
Buber, Martin, 86, 257, 277
Burke, John F., 184, 208
Burke, Peter J., 136
Butler, Judith P., 283

Camus, Albert, 254, 282–83, 290, 298, 304
Cantor, Georg F. L. P., 190
Capra, Fritjof, 257
Carnap, Rudolf, 176
Cassiopeia, 210
Chalmers, Alan F., 11
Chandler, Daniel, 12
Childress, Boyd, 126
Choi, Sungho, 35
Cicero, Marcus T., 5
Clement of Alexandria, 133, 298
Colestock, Richard O., 45
Commoner, Barry, 192
Comte, I. M. Auguste F. X., 64, 301
Confucius (Kong Qiu), 114, 262
Copernicus, Nicolaus, 172
Crystal, David, 25
Cureton, Adam, 72
Cyril of Jerusalem, 116

Darwin, Charles R., 172, 276
David, King, 283
Da Vinci, Leonardo P., xvi, 208, 288
Dawkins, Richard, 236–37, 276
Dawson, John W., 194
Defoe, Daniel, 285
Dennett, Daniel, 39
DePillis, Lydia, 215
Derrida, Jacques, 25, 36, 66, 163, 267
Descartes, René, 7, 13, 24, 247, 255
Dewey, John, 124, 262
Dilthey, Wilhelm, 75
Diogenes of Sinope, 7
Domingos, Pedro, 193, 199–200, 203, 215, 261
Dreyfus, Alfred, 159
Dryden, John, 303

Duhem, Pierre M. M., 186
Duignan, Brian, 302
Durkheim, D. Emile, 110
Dyer, Hugh, 160
Dyer, Jeremy L., 94,

Eddington, Arthur S., 90, 171, 187, 214
Edwards, David Lawrence, 281
Edwards, David Lloyd, 281
Einstein, Albert, 172, 181–82, 184, 185, 190–91, 193, 202, 207–8, 237, 255, 261
Elliott, Charlotte, 60
Epictetus, 64
Etzioni, Amitai, 130
Euclid of Alexandria, 192

Fara, Michael G., 35–36
Ferguson, Kitty, 146, 257
Feyerabend, Paul K., 77, 206
Feynman, Richard P., 141, 172, 182, 193, 257
Fichte, Johann G., 114
Fillingham, Lydia A., 129
Firth, John R., 25
Fletcher, Ronald, 138
Floyd, George P. Jr., 159
Fodor, Jerry A., 22, 23
Foot, Philippa R., 176–77
Foulkes, Paul, xv–xvi, 21, 82, 190, 192, 200, 283
Fox, Melodie J., 34
Fox, Nicholas J., 238
Francis, Pope, 146, 259
Frankel, Charles, 131
Frege, F. L. Gottlob, 26
Fremstedal, Roe, 281
Fry, Stephen J., 30, 275

Galbraith, Deane (Te Rarawa Ngaāpuhi), 54
Galilei, V. G. Galileo de, 184, 192, 255
Gassendi, Pierre, 293
Gehry, Frank O., 90
Gelernter, David H., 95
Geyrhalter, Fabian, 294
Ghazali, Abu H. M. al, 7
Gintis, Herbert, 87

Name Index

Giubilini, Alberto, 138
Glaze, David, 294
Glazer, Nathan, 110
Gödel, Kurt F., 194,
Goodwin, Samuel, 219
Gordon, W. Terrence, 111
Govinda, Lama A., 12
Graham, Daniel W.,
Grandma (Anna M. R.) Moses, 9
Grayling, Anthony C., 201
Grimshaw, Anna, 102
Gropius, Walter A. G., 94

Halcomb, James, 108
Hamawiyyah, Shaykh S. Ibn, 7
Hart, Kevin J., 101, 138, 165
Hartmann, P. Nicolai, 55
Hawking, Stephen W., 8, 63, 75, 173, 213, 216, 276, 279–80, 296
Hayakawa, Samuel I., 16
Hayek, Friedrich A. von, 146
Haytham, H. Ibn (Alhazen) al, 63
Hegel, Georg W. F., 6, 7, 25, 26, 31, 64, 140, 271, 308
Heidegger, Martin, 147, 238
Hemingway, Ernest M., 281
Hepburn, Brian, 182
Heraclitus of Ephesus, 59
Herder, Johann G., 76
Herschel, John F. W., 100
Heshmat, Shahram, 136
Hesse, Hermann K., xvi, 80, 227, 243
Hesse, Mary B., 193
Hitler, Adolf, 295
Hobbes, Thomas, 255
Hodge, Charles, 240
Hofstadter, Douglas R., 32, 33
Hogarth, William, 94
Holbach, Paul-Henri T. d', 275–76
Holder, Rodney D., 194
Hölzel, Adolf R., 94
Hook, Sidney, 286
Horgan, John, 144–45
Horkheimer, Max, 157
Hoyningen-Huene, Paul, 200
Humboldt, F. W. H. Alexander von, 76, 124, 234

Hume, David, 26, 53, 55, 66, 72, 137, 174–75, 194, 217, 226, 233, 234, 259, 297, 300
Hutcheson, Francis, 89
Huxley, Thomas H., 240

Icarus, 210

Jade Emperor (Yuhuang), 86
Jakobson, Roman O., 22, 35
James, William, 7, 56, 69, 182
Jameson, Fredric, 94
Jaspers, Karl T., 26, 64, 96, 163
Jefferson, Thomas, 272–73
Jesus Christ, 243
Joad, Cyril E. M., 43
Johnson, Robert N., 72
Jones, Karen S., 97
Julius Caesar, 243

Kamlah, Wilhelm, 24, 25, 29, 31, 204, 213, 214
Kandinsky, Wassily W., 93
Kant, Immanuel, 7, 9, 14, 36, 56, 72, 85–86, 153, 162, 175, 183, 253, 256
Kapila, 236, 279
Kaplan, Aryeh M. E., 275
Kauffman, Kohler, 287
Kauffman, Stuart A., 279,
Kaunda, Kenneth D., 114
Kennedy, John F., 120
Kenyon, Karl W., 147
Kierkegaard, Søren A., 26, 281, 284
King, Martin L. Jr., 256
Kipling, J. Rudyard, 114
Kirkpatrick, Jeane D., 113
Knox, Ronald J., 16
Korten, David C., 54
Korzybski, Alfred H. S., 230
Kridler, Eugene, 147
Kuhn, Thomas S., 69, 208
Küng, Hans, 5, 162, 175
Kusama, Yayoi, 289
Kuznicki, Jason, 85–86

Ladyman, James, 170
Lamey, Andy, 148
Laozi (Lao-Tzu), 36, 58, 227

Laplace, Pierre–Simon, Marquis de, 276
Laurence, Stephen, 19–20, 29, 49
Lavoisier, Antoine–Laurent de, 180
Lazarus, Richard S., 72
Leary, Timothy F., 259
Leech, Geoffrey N., 42, 43, 46
Leibniz, Gottfried W., 24, 248–49, 255
Lichtenberg, Georg C., 43
Löbner, Sebastian, 29, 80, 81
Locke, John, 19, 46, 47, 113, 115, 252–53, 260, 305
Lorenzen, Paul, 24, 25, 29, 31, 204, 213, 214
Losonsky, Michael, 32
Lotze, Rudolf H., 177
Luntley, Michael, 180
Lyotard, Jean–François, 9, 10, 218
Lyth, David, 184

MacBride, Fraser, 49
Mach, Ernst W. J. W., 56, 183, 235
Machiavelli, Niccolò, 112
MacKinnon, James B., 216
Madison, James Jr., 115
Mandela, Nelson R., 75
Mannan, M. Abdul, 180
Marcuse, Herbert, 219
Margolis, Eric, 19–20, 29, 49
Marx, Karl H., 30, 124, 219, 267
Matthews, Peter, 22, 25
Mauthner, Fritz, 230
Mautner, Thomas, 43, 191
Maxwell, James C., 255
McAleer, Sean, 5
McLuhan, H. Marshall, 111
Melanby, Kenneth, 146
Midgley, Mary B., 204
Montesquieu, Charles–Louis de, 115
Mlodinow, Leonard, 8
Morin, Edgar, 260
Moyn, Samuel A., 138
Mozart, Wolfgang A., 91
Munch, Edvard, 289
Munitz, Milton K., 11

Nagel, Thomas, 248, 292
Needleman, Jacob, 212
Newton, Isaac, 172, 181, 182, 184, 275

Ng, Ruth, 20
Niebuhr, K. P. Reinhold, 245
Nietzsche, Friedrich W., 7, 25, 26, 78, 236
Niobe, 210

O'Connor, John J., 323
Ohm, Georg S., 185, 206, 230
Opp, Karl–Dieter, 178
Oreskes, Naomi, 145
Orzel, Chad, 191
Osborne, John, 94
Osborne, Richard, 5, 9, 10, 298
Osgood, Charles E., 227
Osiander, Andreas, 182
Ovid (Publius O. Naso), 73

Paech, Niko, 214
Paine, Thomas, 123
Pang Cong, 244
Parfit, Derek A., 91
Pascal, Blaise, 72, 373, 302, 304
Pasteur, Louis, 211
Paton, Alan S., 259
Paul of Tarsus, 131
Peirce, Charles S., 46, 64, 182, 199
Phaethon, 210
Piaf, Édith (Édith Giovanna Gassion), 91
Picasso, Pablo R., 93
Picquart, Marie–Georges, 159
Pinker, Steven A., 118
Plato, 5, 7, 24, 25, 76, 89, 92, 109–10, 153, 186, 234, 259, 262
Plotinus, 257
Poincaré, J. Henri, 208
Pol Pot (Saloth Sâr), 295
Polanyi, Michael (*Polányi* Mihály), 199, 267
Pollock, P. Jackson, 90
Popper, Karl R., 234
Preston, Douglas, 211
Prinz, Jesse J., 250
Putin, Vladimir V., 37, 80
Putnam, Hilary W., 291
Pylyshyn, Zenon W., 22, 23

Quine, Willard van O., 72
Quinton, Anthony M., 227

Name Index

Redaelli, Simone, 181
Rettler, Bradley, 15
Reuell, Peter, 110
Rico, Luis (Romero, Luis R.), 261
Roberts, Peter R., 93
Robertson, Edmund F., 206
Rolston, Holmes III, 143, 144
Roosevelt, Franklin D., 114
Rosen, Gideon, 15
Ross, William D., 79
Rousseau, Jean-Jaques, 7, 87
Rowe, Jonathan, 16
Rumi, Jalal A. M., 88
Ruskin, John, 110
Russell, Bertrand A. W., 7, 10, 46, 53, 172, 175, 182, 202, 233, 234
Rutherford, Ernest, 183-84

Sadiq, Jafar I. M. al, 161
Saint-Simon, Claude-Henri de, 125
Saito, Yuriko, 90
Salmoneus, 210
Sánchez, Francisco, 63
Sander, Emmanuel, 32, 33
Sapir, Edward, 100
Sartre, Jean-Paul C. A., 79, 80, 151, 238
Saussure, Ferdinand de, 19, 95, 198
Sautet, Marc, 144
Savigny, Friedrich C. von, 130-31
Sayer, Andrew, 83
Scarborough, Mirjam R., 85
Scarborough, Thomas O., 207
Schaeffer, Francis A., xv, 9, 76, 164, 239, 245
Schiller, J.C. Friedrich, 86,
Schlick, F.A. Moritz, 8
Schoenberg (Schönberg), Arnold, 94
Schopenhauer, Arthur, 7, 302
Schrag, Calvin O., 137
Schrödinger, Erwin R. J. A., 172
Schwartz, Laurent-Moïse, 209
Schweitzer, L. P. Albert, 114
Seiler, Robert M., 157
Sherwood, S. William, 34
Sia, Paolo di, 95
Sidgwick, Henry, 215
Simon, Yves R. M., 127, 130, 132
Sloman, Aaron, 15, 192

Smith, Adam, 127
Soanes, Catherine, 171
Socrates, 5, 78, 157, 262, 285
Solomon, King, 296
Sontag, Susan, 240
Spinoza, Baruch (Bento) de, 111
Stahl, Titus, 284
Steiner, F. George, 22
Stravinsky, Igor F., 90
Strawson, Peter F. , 86
Suzuki, Daisetsu (Daisetz) T., 255

Tagore, Rabindranath, 69, 163, 291
Tegmark, Max E., 213
Teilhard de Chardin, Pierre, 64, 281, 287
Tereus, 210
Thales of Miletus, 7, 259
Thomas, Christelle, 94
Thomas, Norman L., 181, 182, 194, 229
Thompson, Mel, 13, 54, 182, 207, 216, 226, 282
Tillich, Paul J., 37, 102, 216, 281, 284, 295
Toulmin, Stephen E., 48, 172, 187, 242
Travis, Stephen H., 283, 284

Ullian, Joseph S., 72

Vaihinger, Hans, 181
Varzi, Achille C., 228, 229
Verwoerd, Elizabeth (Betsie), 75
Vesey, Godfrey N. A., xv-xvi, 21, 82, 190, 192, 200, 283
Voltaire (Arouet, François-Marie), 7, 277

Wagner, Roy, 103
Ward, Graham J., 293
Watson, John B., 8
Weatherford, Jack M., 124
Webb, Richard, 185, 194, 208
Weber, Maximilian (Max) K. E., 109, 175, 218
Weil, Simone A., 6, 256
Weinreich, Peter, 135
Wellman, Carl, 138
Werbock, Jeffrey, 325
Whitehead, Alfred N., 43, 193, 267, 297
Whorf, Benjamin L., 37, 100
Williams, Raymond H., 135

Wilson, Edward O., 149
Wilson, Kelly G., 252
Wittgenstein, Ludwig J. J., 7, 10, 20, 41, 54, 55, 89, 90, 175, 221, 226, 233, 267, 300
Wolchover, Natalie, 194, 280
Wood, W. Jay, 246
Wren, Christopher, 91

Yannas, Ioannis V., 184, 208

Zangwill, Nick, 89
Zedong, Mao (Chairman Mao), 75
Zeno of Elea, 234

www.ingramcontent.com/pod-product-compliance
Lightning Source LLC
Chambersburg PA
CBHW050330230426
43663CB00010B/1801